Competency Based Mathematics

for

Secondary Schools

Book 2

(MODULES 5 TO 9)

Nji Emmanuel Ndi
GBHS Mankon - Bamenda
North West Region Cameroon
Tel: (+237) 676 684 050
Email: manuelndike@gmail.com

First Edition

Printed by CreateSpace, an Amazon.com Company

EStore address: www.CreateSpace.com/6812338

Available from Amazon.com, CreateSpace.com, and other retail outlets

Available on Kindle and other retail outlets

Books by Nji Emmanuel Ndi

Complete Ordinary Level Mathematics Passport
Rudiments of Ordinary Level Mathematics
Advanced Level Pure Mathematics Key Facts
Competency Based Mathematics for Secondary Schools Book 1
Competency Based Mathematics for Secondary Schools Book 2
Competency Based Mathematics for Secondary Schools Book 3
Competency Based Mathematics for Secondary Schools Book 4
Competency Based Mathematics for Secondary Schools Book 5

Copyright © 2017 Nji Emmanuel Ndi
All rights reserved.
ISBN-10: 1541255593
ISBN-13: 978-1541255593

DEDICATION

Dedicated to all emerging and emergent Societies

Competency Base Mathematics for Secondary Schools Book 2

Table of Contents

DEDICATION I
   Acknowledgement vi
   How to Use this Book vii
   Notation Used in this Book ix

MODULE 5: 1

NUMBERS, FUNDAMENTAL OPERATIONS AND RELATIONSHIPS IN THE SET OF NUMBERS 1

TOPIC 1 : THE SET OF INTEGERS 2

   1.1   Review and Revision 3
   1.2   Whole Number Powers 3
   1.3   Multiplication Law of Indices 3
   1.4   Division Law of Indices 4
   1.5   The Zero Index Law 5
   1.6   The Negative Index Law 6
   1.7   Power of an Exponential Number Law 7

TOPIC 2 : NUMBER PATTERNS 10

   2.1   Squaring and Cubing Exponential Numbers 11
   2.2   Square Roots and Cube Roots of Exponential Numbers 12
   2.3   Square Roots and Cube Roots of Whole Numbers 12
   2.4   Square Roots and Cube Roots of Fractions 13
   2.5   Elementary Sequence and Patterns 15

TOPIC 3 : FRACTIONS AND DECIMALS 20

   3.1   Rounding Down and Rounding Up 22
   3.2   Estimations and Approximations 23
   3.3   Decimal Places 25
   3.4   Significant Figures 25
   3.5   Standard Form 27
   3.6   Application of Fractions and Decimals to Real Life 29

TOPIC 4 : ARITHMETIC PROCESSES 34

   4.1   Variation in Real Life 35
   4.2   Direct Variation 36
   4.3   Using the Unitary Method to Solve Proportions 36
   4.4   Inverse Variation 38
   4.5   Profit and Loss 39
   4.6   Percentage Profit And Loss 40
   4.7   Basic Terms Related to Interest 42
   4.8   Simple Interest 42
   4.9   Compound Interest 44
   4.10   The Compound Interest Formula 45

4.11	COMPOUND INTEREST WITH VARYING PRINCIPAL	46
4.12	CURRENCY EXCHANGE- DECIMAL CURRENCY	48

TOPIC 5 : REAL NUMBERS ... 52

5.1	RADICALS	53
5.2	IRRATIONAL NUMBERS $\mathbb{Q}'$	53
5.3	REAL NUMBERS, $\mathbb{R}$	54
5.4	THE REAL NUMBER LINE	54
5.5	RELATIONSHIP BETWEEN SETS OF NUMBERS	55
5.6	INTERVALS	57
5.7	ORDERING	58
5.8	REPRESENTATION OF INTERVALS	58

MODULE 6: ... 65

INTRODUCTION TO PLANE GEOMETRY ... 65

TOPIC 6 : DISTANCES ... 66

6.1	HORIZONTAL OR VERTICAL DISTANCE BETWEEN TWO POINTS	67
6.2	MIDPOINT AND MEDIATOR OF A LINE SEGMENT	69
6.3	CONSTRUCTING THE MEDIATOR OF A LINE SEGMENT	69

TOPIC 7 : ANGLES ... 74

7.1	ANGLES	75
7.2	PARALLEL LINES AND TRANSVERSALS	75
7.3	INTERIOR AND EXTERIOR ANGLES OF POLYGONS	78
7.4	SUM OF ANGLES OF A POLYGON	78
7.5	POLYGON THEOREMS	80

TOPIC 8 : TRIANGLES ... 87

8.1	STANDARD NOTATION FOR TRIANGLES	88
8.2	INTERIOR AND EXTERIOR ANGLES OF TRIANGLES	88
8.3	CHASLES' THEOREM	89
8.4	THE RIGHT-ANGLED TRIANGLE	90
8.5	THE PYTHAGORAS THEOREM	92
8.6	CONGRUENT TRIANGLES	95
8.7	CONDITIONS FOR TRIANGLES TO BE CONGRUENT	96

MODULE 7: ... 102

SOLID FIGURES ... 102

TOPIC 9 : PRISMS AND CYLINDERS ... 103

9.1	OBSERVATION AND DESCRIPTION OF PRISMS	104
9.2	EDGES, VERTICES, FACES OF SOLID FIGURES	105
9.3	NETS OF PRISMS	105
9.4	SURFACE AREA AND VOLUME OF PRISMS	106
9.5	SURFACE AREA AND VOLUME OF CUBES	108
9.6	SURFACE AREA AND VOLUME OF CYLINDERS	109

TOPIC 10 : PYRAMIDS 114

    10.1   Observation and Description of Pyramids 115
    10.2   Naming Pyramids 115
    10.3   Polyhedrons 116
    10.4   Mensuration of Pyramids 117
    10.5   Volume of Pyramids 117

TOPIC 11 : SCALES AND SIMILARITY 122

    11.1   Plans and Maps 123
    11.2   Similar Figures 124
    11.3   Congruent Figures 128

MODULE 8: 137

ELEMENTARY STATISTICS AND PROBABILITY 137

TOPIC 12 : REPRESENTATION OF DISCRETE DATA 138

    12.1   Discrete and Continuous Data 139
    12.2   Frequency Distribution Table for Discrete Data 140
    12.3   Statistical Graphs 141

TOPIC 13 : MEASURES OF CENTRAL TENDENCIES 147

    13.1   Notion of Measures of Central Tendencies 148
    13.2   Mode 148
    13.3   Median 148
    13.4   Arithmetic Mean (Average or Mean) 149

TOPIC 14 : ELEMENTARY PROBABILITY 154

    14.1   The Concept of Probability 155
    14.2   Some Basic Probability Terminology 155
    14.3   Probability as a Number 157
    14.4   Equiprobable Outcomes 157
    14.5   Standard Definition of Probability 158
    14.6   Complementary Events 160

MODULE 9: 164

BASIC ALGEBRA 164

TOPIC 15 : ALGEBRAIC EXPRESSIONS 165

    15.1   Symbolic Expressions 166
    15.2   Variables 166
    15.3   Algebraic Sentences (Expressions) 166
    15.4   From English to Algebra 167
    15.5   Variable Substitution 168
    15.6   Unknowns and Constants 169
    15.7   Terms and Coefficients 169

15.8	LIKE AND UNLIKE TERMS	170
15.9	ALGEBRAIC RULES	170
15.10	COMBINING LIKE TERMS	171
15.11	MULTIPLICATION AND DIVISION OF TERMS	172
15.12	EXPANSIONS	173
15.13	FACTORISATION	175

TOPIC 16 : SIMPLE LINEAR EQUATIONS AND INEQUALITIES ... 178

16.1	THE CONCEPT OF AN EQUATION	179
16.2	SIMPLE (LINEAR) EQUATIONS	179
16.3	DIFFERENCE BETWEEN EQUATIONS AND EXPRESSIONS	180
16.4	ADDITIVE INVERSES	181
16.5	INVERSE OPERATIONS	181
16.6	SOLVING SIMPLE LINEAR EQUATIONS	183
16.7	ONE STEP SIMPLE LINEAR EQUATIONS	185
16.8	TWO-STEP SIMPLE LINEAR EQUATIONS	187
16.9	MULTI-STEP SIMPLE LINEAR EQUATIONS	188
16.10	SIMPLE LINEAR EQUATIONS INVOLVING FRACTIONS AND DECIMALS	189
16.11	WORD PROBLEMS ON SIMPLE LINEAR EQUATIONS	191
16.12	INEQUALITIES AND INEQUATIONS	193
16.13	REPRESENTATION OF INEQUALITIES	194
16.14	BUILDING UP INEQUATIONS	194
16.15	LAWS OF INEQUALITIES	196
16.16	SOLVING INEQUATIONS	198

ANSWERS TO STRUCTURAL EXERCISES ... 204

Acknowledgement

My deepest gratitude goes to God Almighty for the inspiration and for the strength.

Many thanks go to Mme. Mbuameh Daisy and Mr. Mburubah Walters for their critical proof reading of the typescript and for offering very useful suggestions which went a long way to reshape the work, the North West Regional Pedagogic Inspector for Mathematics Mr. Nfor Samuel Ndi who preview the initial manuscript and gave ample advice, which went a long way to reshape the document. I heartily thank the Former North West Regional Pedagogic Inspector for Mathematics Mr. Nji Samuel Tatah who made a very commendable effort to edit the Mathematics content of the book. I cannot forget the last minute encouragements and advice which the National inspector of Mathematics Mme Babila Emilia inspired me with. I equally pay much tribute to my students on which this material was tested. I cannot end here without thanking my sweet heart Nji Irene Nfih and my Children who encouraged and supported me in one way or the other during the course of the work.

Many thanks go to the WAEC and the CGCE Board for allowing their past questions to be used directly or indirectly.

Nji Emmanuel Ndi

G.B.H.S. Mankon, Bamenda

North West Region

Cameroon

TEL: (+237)76684050

E-mail: manuelndike@gmail.com

How to Use this Book

This book is written in a very special way with different sections boxed and represented by special symbols as follows.

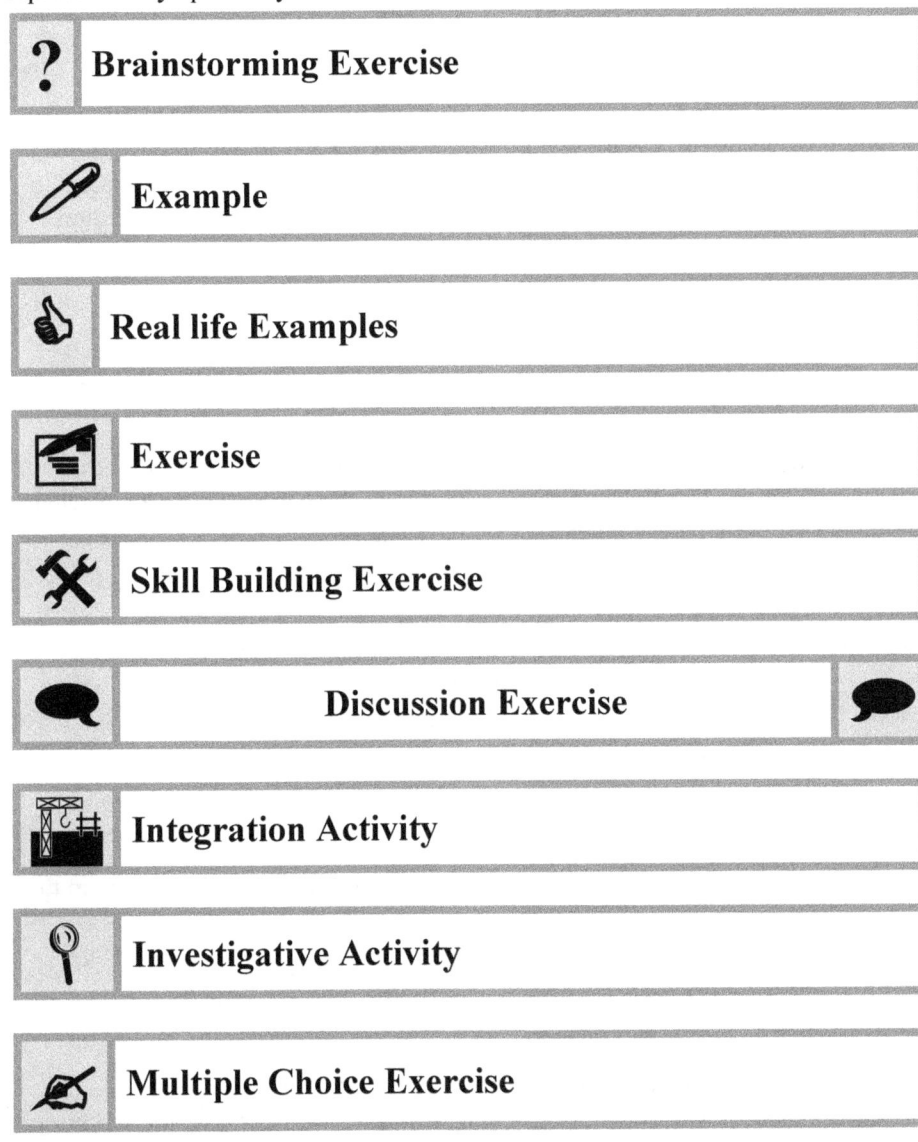

👥 Group Activity

The various sections represented by different symbols are out to facilitate navigation through the book. By investing enough time and energy in each section both students and teachers will realize that their speed and understanding will be greatly enhanced.

The brain storming exercises are aimed at provoking and invoking the learners' minds to prepare them for the task at hand. The teacher is highly encouraged to orally question the students during lessons using questions under this section.

The investigative exercises are meant to give the learner ample opportunity to experiment and self-discover facts and concepts and develop methods and skills without being told.

The group activities and discussion exercises are aimed at developing a team spirit in the learners.

Many well designed examples are vividly used and solved to facilitate the learner's understanding by showing the necessary steps required for a particular solution. There are a good number of real life examples which point out the application of the subject matter in real life situations. The student is advised to study these examples very carefully.

There are many well graded exercises and skill building exercises to test the level of understanding of the learner and to facilitate skill development in the learner. The student is advised to attempt all the questions as each question may have its own technique.

Many integration activities have been designed to unify groups of sub topics, topics or modules in some cases.

Where necessary review exercises have been given to help the learner retain the skills acquired in the earlier sections.

Finally each topic ends with a good number of multiple choice questions. In each question only one of the alternatives is correct. Write down the letter corresponding to the correct answer.

For greatest achievement, the learner is advised to study regularly what he does not know and work without fear of making mistakes whether with the teacher or during group work.

By consistently and systematically going through this course as instructed, the learner will be overwhelmed with the competencies acquired at each level and at the end of the course.

How to Use this Book

Notation Used in this Book

$+$	Addition
$-$	Subtraction
$\times$	Multiplication
$\div$	Division
$\%$	Percentage
$\{\ldots\}$	The set of elements…or the unordered list with elements…
$n(A)$	The number of elements in set A
$\{x:\quad\}$	The set of all x such that
$\in$	Is an element of …
$\notin$	Is not an element of …
$\{\ \}$ or $\emptyset$	The empty set.
$\mathscr{E}$	The universal set.
$\cup$	The union of …
$\cap$	The intersection of…
$\subseteq$	Is a subset of …
$\subset$	Is a proper subset of …
$\mathbb{Z}$	The set of integers, $\{0, \pm1, \pm2, \pm3, \pm4, \ldots\}$
$\mathbb{N}$	The set of all positive integers and zero, $\{0,1,2,3,4,\ldots\}$
$\mathbb{Z}^+$	The set of positive integers $\{+1, +2, +3, +4 \ldots\}$
$\mathbb{Q}$	The set of rational numbers
$\mathbb{Q}^+$	The set of positive rational numbers
$\mathbb{R}$	The set of all real numbers $\{x: x \in \mathbb{R}\}$
$\mathbb{R}^+$	The set of all positive real numbers $\{x \in \mathbb{R}: x > 0\}$
$=$	Is equal to
$\neq$	Is not equal to
$\approx$	Is approximately equal to
$<$	Is less than
$>$	Is greater than
$\not<$	Is not less than
$\not>$	Is not greater than
$\leq$	Is less than or equal to
$\geq$	Is greater than or equal to

$a < x < b$ or $]a, b[$ or (a, b)	An open interval on the number line
$a \leq x \leq b$ or $[a, b]$	A closed interval on the number line
$\{x : a < x < b\}$	The set of elements x such that a is less than x and x is less than b
x^n	The number x, raised to the power n
$\propto$	Is proportional
$\sqrt{}$	The positive square root
$\sqrt[3]{}$	Cube root
$\perp$	Is perpendicular to
$\parallel$	Is parallel to
$\angle$	Angle
$^\circ$	Degree
$^\circ C$	Degrees Celsius
$^\circ F$	Degrees Fahrenheit
Σ	Summation
$\bar{x}$	Average or mean

Module 5

Numbers, Fundamental Operations and Relationships in the set of Numbers

Family of Situations
Module 5 is an extension of module 1 and at the end of this module; the student is expected to acquire more competencies within the **families of situations** *'Representation, determination of quantities and identification of objects by numbers'*.

Categories of Action
The categories of action for module 5 include:
1. Determination of a number,
2. Reading and writing information using numbers,
3. Verbal interaction on information containing numbers,
4. Estimation and treatment of quantities.

Credit
The module is expected to be covered within 8 weeks teaching 4 hours per week (or within 30 to 32 hours).

Topic 1

THE SET OF INTEGERS

Objectives

At the end of this topic, the learner should be able to:

1. Determine the value of a number expressed using whole number powers.
2. Recognize and state the following laws of indices.
   (i)   The multiplication law of indices.
   (ii) The division law of indices.
   (iii) The zero index law.
   (iv) The negative index law.
   (v)   The power of an exponential number law.
3. Use the above laws of indices to solve simple numerical problems involving indices.

Module 5, Topic 1: The Set $\mathbb{Z}$ of Integers

1.1 Review and Revision

 Review Exercise 1

1. The set of natural numbers is denoted by……………
2. The set of counting numbers is denoted by…………
3. List the first five elements of the sets in 1.
4. Arrange the numbers 77, 98, 79, 76 in increasing order of magnitude.
5. Use the symbols < or > to compare the following.
   (a) 3×4 $2+7$ (b) $18 + 6$ 7
6. Using the symbol < arrange the numbers 63, 56, 74, 68 in increasing order of magnitude.
7. In 5^6, 6 is called ……………, ………….. or ………… and 5 is called ……
8. Write out the following in index form.
   (a) $3 \times 3 \times 3 \times 3 \times 3 \times 3 \times 3 \times 3 \times 3$ (b) $6 \times 6 \times 6 \times 6 \times 6$
9. Write each of the following as a product. (a) 9^2 (b) 5^3
10. Write 10,000,000 in index form to the base 10.
11. Evaluate (a) 3^4 (b) 4^3
12. List the first five (a) odd numbers (b) even numbers.
13. Evaluate (a) $7 \times 9 + 4 \times 10$ b) $12 + 8 \times 6 + 64$
14. List the elements of the set of integers and write down the symbol which denotes this set.

1.2 Whole Number Powers

In section 2.6 of book 1, we saw,

$$5^{3} \quad \leftarrow \text{power, index or exponent}$$
$$\uparrow$$
$$\text{base}$$

5 is called the base and 3 is called the **power, index, exponent** or **logarithm**.

Note!! Any base with exponent 1 is itself.

1.3 Multiplication Law of Indices

 Investigative Activity

1. Expand and evaluate the following leaving your answers in index form.
   (a) $3^5 \times 3^3$ (b) $4^3 \times 4^2$
2. What do you say about the bases in each case in 1(a) and (b)?
3. Add the exponents in 1(a) and (b) and compare your results with your answers.
4. What conclusion do you draw?

From the above investigation, we see that in both (a) and (b) the results could be obtained by adding the indices. Thus,

(a) $3^5 \times 3^3 = 3^{5+3} = 3^8$ (b) $4^3 \times 4^2 = 4^{3+2} = 4^5$

To multiply two numbers written to the same base in index form, keep the base and add the exponents or indices.

 Example

Without expanding, evaluate the following allowing your answers in index form.
(i) $5^2 \times 5^4 \times 5^1 \times 5^3$ (ii) $7^4 \times 7^2 \times 7^3 \times 7^0$

Solution
(i) $5^2 \times 5^4 \times 5^1 \times 5^3 = 5^{2+4+1+3} = 5^{10}$ (ii) $7^4 \times 7^2 \times 7^3 \times 7^0 = 7^{4+2+3+0} = 7^9$

1.4 Division Law of Indices

 Investigative Activity

1. Expand and evaluate the following leaving your answers in index form.
   (a) $3^7 \div 3^5$ (b) $4^5 \div 4^2$
2. What do you say about the bases in each case in 1(a) and (b)?
3. Subtract the exponents in 1(a) and (b) and compare your results with your answers.
4. What conclusion do you draw?

From the above investigation, we see that in both (a) and (b) the results could be obtained by subtracting the indices. Thus,

(a) $3^7 \div 3^5 = 3^{7-5} = 3^2$ (b) $4^5 \div 4^2 = 4^{5-2} = 4^3$

To divide two numbers written in index form to the same base, keep the base and subtract the exponents or indices.

Example

1. Evaluate the following leaving your answer in index form where necessary.

   (a) $7^9 \div 7^4$ (b) $\dfrac{5^9}{5^6}$

   Solution

   (a) $7^9 \div 7^4 = 7^{9-4} = 7^5$ (b) $\dfrac{5^9}{5^6} = 5^{9-6} = 5^3$

2. Simplify leaving your answer in index form where necessary.

   (a) $6^3 \div 6^4 \times 6^2$ (b) $\dfrac{3^5 \times 3^4}{3^6 \times 3^2}$

   Solution

   (a) $6^5 \div 6^4 \times 6^2 = 6^{5-4+2} = 6^3$ (b) $\dfrac{3^5 \times 3^7}{3^6 \times 3^2} = 3^{5+7-6-2} = 3^4$

1.5 The Zero Index Law

 Investigative Activity

1. Expand and evaluate the following leaving your answer in index form where necessary. (a) $3^5 \div 3^5$ (b) $4^3 \div 4^3$ (b) $0^7 \div 0^7$
2. What do you say about the bases in each case in 1(a) and (b)?
3. Expand and evaluate the above completely by cancelling factors.
4. What conclusion do you draw concerning your answers?

From the above investigation, we see that:

Any base other than 0 raised to the power zero is 1.

 Example

Evaluate the following.

(a) $11^6 \div 11^6$ (b) $\dfrac{2^5 \times 2^3}{2^2 \times 2^6}$

Solution

(a) $11^6 \div 11^6 = 11^{6-6} = 11^0 = 1$ (b) $\dfrac{2^5 \times 2^3}{2^2 \times 2^6} = 2^{5+3-2-6} = 2^0 = 1$

1.6 The Negative Index Law

 Investigative Activity

1. Use the division law to evaluate the following leaving your answer in index form. (a) $3^3 \div 3^7$ (b) $11^4 \div 11^6$
2. Expand and evaluate the above by cancelling factors leaving your answer in index form.
3. What conclusion do you draw concerning your answers?

From the above investigation, we see that:

Any number raised to a negative index is the same as the reciprocal of the same number raised to the corresponding positive index.

 Example

Evaluate the following.

(a) $\left(\dfrac{5}{2}\right)^{-1}$ (b) $\dfrac{5^7 \times 5^{-3}}{5^{-2} \times 5^4}$

Solution

(a) $\left(\dfrac{5}{2}\right)^{-1} = \left(\dfrac{2}{5}\right)^1 = \dfrac{2}{5}$ (b) $\dfrac{5^7 \times 5^{-3}}{5^{-2} \times 5^4} = \dfrac{5^{7+(-3)}}{5^{-2+4}} = \dfrac{5^4}{5^2} = 5^{4-2} = 5^2 = 25$

 Skill Building Exercise 1:1

Evaluate completely each of the following.
1. 2^6
2. 8^2
3. $2^3 \times 3^2$
4. $3^3 \times 5^2$
5. $3^3 \times 3^2 \times 3^0$
6. $5^2 \times 2 \times 5^1$
7. $4 \times 3 \times 3^3$
8. $13^4 \div 13^6$
9. $16 \times 3^5 \div (8 \times 3^3)$
10. $25 \times 3^5 \div (5 \times 3^4)$
11. $\dfrac{2^3 \times 3^4 \times 4^7}{2^2 \times 3^2 \times 4^5}$
12. $\dfrac{10 \times 4^6}{5 \times 4^3}$
13. $\dfrac{2 \times 5^2 \times 7}{3 \times 5 \times 7^2}$
14. $\dfrac{12 \times 5^{-2}}{9 \times 5^{-3} \times 4}$
15. $\dfrac{-5 \times 7^2}{7^{-1}}$

16. $\dfrac{5 \times 4^2 \times 3^2 \times 2^{-1}}{15 \times 4^{-2} \times 32}$

17. $\dfrac{5^3 \times 7^0 \times 36 \times 2^{-4}}{4^{-1} \times 5^4 \times 9^2}$

18. $\dfrac{-5^2 \times 3}{2 \times (-5)^2}$

19. $\dfrac{3^{-1} \times 2^3 \times 5^{-1}}{2^{-1} \times 5^2}$

20. $\dfrac{5^3 \times 5^0}{25}$

21. $5 \times 7^2 \times 3^{-1} \times 7^{-1} \div (5^{-1} \times 7)$

22. $5\dfrac{2}{5} \times \left(\dfrac{2}{3}\right)^2 \div \left(1\dfrac{1}{2}\right)^{-1}$

23. $6^{-3} \times 2^5 \times 3^3$

24. $\dfrac{1}{3^{5 \times 2}} \times 9^{2-1} \times 27^{2+1}$

1.7 Power of an Exponential Number Law

Investigative Activity

1. Expand and evaluate the following leaving your answers in index form.

   (a) $\left(3^2\right)^3$ (b) $\left(2^3\right)^4$

2. Multiply the exponents in 1(a) and (b) and compare your results with your answers.
3. What conclusion do you draw?

From the above investigation, we see that:

We can evaluate the power of an exponential number by keeping the base and multiplying the powers.

Example

Evaluate the following leaving your answers in index form.

(a) $\left(5^3\right)^7$ (b) $\left(4^{-6}\right)^3$ (c) $\left(3^{-5}\right)^{-3}$

Solution

(a) $\left(5^3\right)^7 = 5^{21}$ (b) $\left(4^{-6}\right)^3 = 4^{-18}$ (c) $\left(3^{-5}\right)^{-3} = 3^{15}$

Skill Building Exercise 1:2

Evaluate each of the following leaving your answer in index form where necessary.
1. $(2^6)^4$
2. $(3^3)^{-4}$
3. $(5^{17})^0$
4. $(13^{-2})^{-5}$
5. $(6^0)^{18}$
6. $(-2^6)^4$
7. $(-2^3)^5$
8. $(-3^{-2})^4$
9. $(-3^{-2})^{-4}$
10. $\dfrac{(2^3)^4 \times 4^7}{2^2 \times 3^2 \times (4^5)^2}$
11. $\dfrac{12 \times (5^3)^{-2}}{9 \times 5^{-3} \times 4}$
12. $\dfrac{-5 \times 7^2}{(-7^3)^{-1}}$

 Multiple Choice Exercise 1

1. The value of $2^2 + 3^3$ is:
   [A] 13   [B] 25   [C] 31   [D] 36
2. 5^4 has a value of:
   [A] 9   [B] 20   [C] 125   [D] 625
3. The value of $2^0 - 2^{-2}$ is:
   [A] 1   [B] $-\dfrac{1}{4}$ [C] $\dfrac{3}{4}$ [D] $\dfrac{1}{4}$
4. $\dfrac{2^3 \times 2^4}{2^2}$ is equal to:
   [A] 2^4 [B] 2^3 [C] 2^2 [D] 2^5
5. The value of $2^5 \times 2^{-3} \times 3^0$ is:
   [A] 20   [B] 2   [C] 4   [D] 10
6. $5 \times 5 \times 5 \times 5 \times 5 \times 5 \times 5 \times 5$ in exponential form is:
   [A] 5×8 [B] 5^8 [C] 55^8 [D] 8^5
7. $5^2(5 - 6^2)$ is equal to:
   [A] −775   [B] 89   [C] 25   [D] −70
8. -7^4 is the same as:
   [A] −3   [B] 2401   [C] −28   [D] −2401
9. $64 \times 4^2 - 3 \times 2^2 =$
   [A] 1031   [B] 4084   [C] 1036   [D] 1012
10. When evaluated $(-4 - (-2))^2 + (-2)^2$ gives:
    [A] −4   [B] −8   [C] 8   [D] 4
11. On simplification $\dfrac{3^2 \times 2^3}{3 \times 2^5}$ gives:
    [A] 1   [B] $\dfrac{1}{5}$ [C] $\dfrac{3}{4}$ [D] $\dfrac{1}{4}$
12. As a single exponent $2^2 \times 2^8$ is:
    [A] 4^{10} [B] 2^{10} [C] 4^{16} [D] 2^{16}
13. The simplified form of $7^5 \times 7^6$ is:
    [A] 49^{30} [B] 7^{30} [C] 49^{11} [D] 7^{11}

14. $\dfrac{144^{14}}{144^2}$ as a single exponent is:

    [A] 144^{16} [B] 144^{12} [C] 144^{28} [D] 144^7

15. $(-1)^0$ is equal to:
    [A] 0 [B] 1 [C] -1 [D] 1^0

16. The quotient $\dfrac{2^8}{2^7}$ is equal to:
    [A] 2^1 [B] 2^{-1} [C] 2^{56} [D] 2^{15}

17. 14^{-1} is certainly:

    [A] $\dfrac{1}{14^4}$ [B] -56 [C] $\dfrac{1}{14^{-4}}$ [D] $\dfrac{1}{14}$

18. The result of squaring the number 6 is:
    [A] 12 [B] 26 [C] 36 [D] 62

19. The smallest number by which we can multiply $3^2 \times 5$ to give a perfect square is:
    [A] 5 [B] 6 [C] 15 [D] 25

20. The least number which multiplies 54 to make a perfect square is:
    [A] 3 [B] 4 [C] 6 [D] 8

Topic 2

NUMBER PATTERNS

Objectives

At the end of this topic, the learner should be able to:

1. Use prime factorization to find the square root of a given natural number by dividing the exponents by 2.
2. Use prime factorization to find the cube root of a given natural number by dividing the exponents by 3.
3. Determine if a set of numbers form a sequence.
4. Determine the rule upon which a sequence is constructed and use it to list the next few terms of the sequence.

Module 5, Topic 2: Number Patterns

2.1 Squaring and Cubing Exponential Numbers

 Brainstorming Exercise

Explain giving two examples in each case the meaning of
(a) Squaring a number.   (b) Cubing a number.

 Investigative Activity

1. By using your ideas in the brainstorming activity above evaluate the following leaving your answers in index form.

    (a) $\left(3^4\right)^2$ (b) $\left(3^4\right)^3$ (c) $\left(5^6\right)^2$ (d) $\left(5^6\right)^3$

2. In 1(a) and (c) double the exponents and compare your results with your answers.
3. What conclusion do you draw?
4. In 1(b) and (d) triple the exponents and compare your results with your answers.
5. What conclusion do you draw?

From the above investigation, we can see that squaring a number doubles its exponent and cubing a number triples its exponent.

Example

Simplify the leaving your answer in index form.
(i) $(11^7)^2$ (ii) $(17^4)^3$

Solution
(i) $(11^7)^2 = 11^{14}$ (ii) $(17^4)^3 = 17^{12}$

2.2 Square Roots and Cube Roots of Exponential Numbers

? Brainstorming Exercise

1. What is the opposite of squaring?
2. What is the opposite of cubing?
3. How can you use your ideas in (1), (2) and the previous investigation to find the square root of a number written in exponential form?
4. How can you use your ideas in (1), (2) and the previous investigation to find the cube root of a number written in exponential form?

The above suggest that;

To find the square root or cube root of a number written in exponential form, divide the exponent by 2 or 3 respectively.

Example

Find the square root of (i) 3^6 (ii) 2^8

Solution

(i) $\sqrt{3^6} = 3^{6 \div 2} = 3^3 = 3 \times 3 \times 3 = 27$ (ii) $\sqrt{2^8} = 2^{8 \div 2} = 2^4 = 2 \times 2 \times 2 \times 2 = 16$

2.3 Square Roots and Cube Roots of Whole Numbers

? Brainstorming Exercise

How can you use your ideas in the previous brainstorming exercise to find the square root or cube root of a composite whole number?

We can find the square root or cube root of a composite whole number by first expressing it as a product of its prime factors and the dividing the power by 2

Module 5, Topic 2: Number Patterns

or 3 respectively.

Example

Evaluate each of the following
(a) $\sqrt{196}$ (b) $\sqrt{900}$ (c) $\sqrt[3]{1000}$ (d) $\sqrt[3]{343}$

Solution
(a) $196 = 2^2 \times 7^2 \Rightarrow \sqrt{196} = 2^{2 \div 2} \times 7^{2 \div 2} = 2 \times 7 = 14$
(b) $900 = 2^2 \times 3^2 \times 5^2 \Rightarrow \sqrt{900} = 2^{2 \div 2} \times 3^{2 \div 2} \times 5^{2 \div 2} = 2 \times 3 \times 5 = 30$
(c) $1000 = 8 \times 125 = 2^3 \times 5^3 \Rightarrow \sqrt[3]{1000} = 2^{3 \div 3} \times 5^{3 \div 3} = 2 \times 5 = 10$
(d) $343 = 7^3 \Rightarrow \sqrt[3]{343} = 7^{3 \div 3} = 7$

Exercise 2:1

1. Square each of the following leaving your answer in exponential form.
   (a) 13^4 (b) 31^3 (c) 17^5 (d) 20^2
2. Evaluate each of the following leaving your answer in exponential form
   (a) $(4^3)^2$ (b) $(7^4)^3$ (c) $(1^5)^3$ (d) $(0^6)^3$
3. Find the square root of each of the following
   (a) 729 (b) 625 (c) 8100 (d) 4096
4. Evaluate each of the following
   (a) $\sqrt{576}$ (b) $\sqrt{1024}$ (c) $\sqrt{1600}$ (d) $\sqrt{1225}$
5. Evaluate each of the following
   (a) $\sqrt[3]{512}$ (b) $\sqrt[3]{1728}$ (c) $\sqrt[3]{5832}$ (d) $\sqrt[3]{2744}$
6. What is the smallest number by which $(5^3) \times 4$ can be multiplied to give a perfect square?

2.4 Square Roots and Cube Roots of Fractions

Investigative Activity

1. Use a calculator to evaluate the following.
   (a) $\sqrt{\dfrac{64}{81}}$ (b) $\sqrt{1\dfrac{29}{196}}$ (c) $\sqrt[3]{\dfrac{343}{729}}$ (d) $\sqrt[3]{2\dfrac{307}{512}}$
2. Evaluate the above by first finding the square root or cube root of both the numerator and denominator. In the cases of the mixed numbers, first convert the

> mixed numbers to an improper fraction.
> 3. Compare your results in 1 and 2.
> 4. What conclusion do you draw?

From the above investigation we see that:

1. To find the square root of a fraction, divide the square root of the numerator by the square root of the denominator.
2. To find the cube root of a fraction, divide the cube root of the numerator by the cube root of the denominator.
3. If the fraction is a mixed number, first convert it to an improper fraction.

Example

1. Evaluate (a) $\sqrt{\dfrac{9}{25}}$ (b) $\sqrt{\dfrac{49}{64}}$

   Solution

   (a) $\sqrt{\dfrac{9}{25}} = \dfrac{\sqrt{9}}{\sqrt{25}} = \dfrac{3}{5}$ (b) $\sqrt{\dfrac{49}{64}} = \dfrac{\sqrt{49}}{\sqrt{64}} = \dfrac{7}{8}$

2. Evaluate (a) $\sqrt[3]{\dfrac{64}{125}}$ (b) $\sqrt[3]{\dfrac{343}{216}}$

   Solution

   (a) $\sqrt[3]{\dfrac{64}{125}} = \dfrac{\sqrt[3]{64}}{\sqrt[3]{125}} = \dfrac{4}{5}$ (b) $\sqrt[3]{\dfrac{343}{216}} = \dfrac{\sqrt[3]{343}}{\sqrt[3]{216}} = \dfrac{7}{6}$

4. Evaluate (a) $\sqrt{6\dfrac{1}{4}}$ (b) $\sqrt[3]{3\dfrac{3}{8}}$

   Solution

   (a) $\sqrt{6\dfrac{1}{4}} = \sqrt{\dfrac{25}{4}} = \dfrac{5}{2}$ (b) $\sqrt[3]{3\dfrac{3}{8}} = \sqrt[3]{\dfrac{27}{8}} = \dfrac{3}{2}$

Exercise 2:2

Evaluate the following.

1. $\sqrt{\dfrac{4}{9}}$ 2. $\sqrt{\dfrac{36}{49}}$ 3. $\sqrt{2\dfrac{1}{4}}$ 4. $\sqrt{\dfrac{9}{16}}$ 5. $\sqrt[3]{\dfrac{27}{343}}$ 6. $\sqrt[3]{\dfrac{125}{27}}$

7. $\sqrt{\dfrac{49}{64}}$ 8. $\sqrt{2\dfrac{7}{9}}$ 9. $\sqrt{\dfrac{81}{4}}$ 10. $\sqrt{7\dfrac{1}{9}}$ 11. $\sqrt{2\dfrac{46}{49}}$ 12. $\sqrt[3]{3\dfrac{3}{8}}$

2.5 Elementary Sequence and Patterns

The Concept of a number Sequence

In book 1, we saw that we can classify numbers as square numbers, rectangular numbers or triangular numbers depending on the figure they form when arranged as dots. As a reminder

Let T = All triangular numbers

S = All square numbers

R = All rectangular numbers

Then,

$T = \{0, 1, 3, 6, 10, 15, 21, 28, \ldots\}$

$S = \{0, 1, 4, 9, 16, 25, 36, \ldots\}$

$R = \{0, 1, 2, 6, 8, 9, 10, 12, 14, 15, \ldots\}$

In addition to square numbers, rectangular numbers and triangular numbers, we can also arrange numbers following many different patterns and rules. Some common rules are:

(a) Adding a given constant number to the preceding one to obtain the next one.

(b) Multiplying a given constant number by the preceding one to obtain the next one.

When this is done an ordered succession of numbers called a number sequence is

Competency Base Mathematics for Secondary Schools Book 2

obtained. Thus,

A **number sequence** is a group of numbers given in a specified order. A sequence may or may not have any definite rule. Examples of number sequences are:

(a) 1, 5, 9, 13...
(b) 1, 5, 25, 125...
(c) 1, 2, 4, 7...
(d) 62, 59, 57, 54, 52...
(e) 5, 7, 13, 16...

Each of the numbers in a sequence is called a **term**.

A sequence which has a last term is called a **finite sequence**. On the other hand a sequence which continues indefinitely is called an **infinite sequence**.

Sequence recognition

For sequences that have a definite rule, we can find the rule and write the next few terms.

Example

Determine the rule and use it to write down the next three terms in each of the following sequences.
(a) 1, 5, 9, 13... (b) 1, 5, 25, 125... (c) 1,2,4,7... (d) 62, 59, 57, 54, 52...

Solution
(a) By observation the rule for this sequence is 'add 4 to a term to obtain the next term'.

$$1 \xrightarrow{+4} 5 \xrightarrow{+4} 9 \xrightarrow{+4} 13 \xrightarrow{+4} 17$$

Therefore the next three terms are;
$13 + 4 = 17, 17 + 4 = 21$ and $21 + 4 = 25$.
So the sequence is 1, 5, 9, 13, 17, 21, 25...

(b) By observation the rule is 'multiply a term by 5 to obtain the next term'

$$1 \xrightarrow{\times 5} 5 \xrightarrow{\times 5} 25 \xrightarrow{\times 5} 125 \xrightarrow{\times 5} 625$$

Therefore the next three terms are;
$125 \times 5 = 625, 625 \times 5 = 3125$ and $3125 \times 5 = 15625$.
So the sequence is 1, 5, 25, 125, 625, 3125, 15625...

(c) By observation the rule is add 1, 2, 3 etc. to the first, second, third terms

Module 5, Topic 2: Number Patterns

respectively.

$$1 \xrightarrow{+1} 2 \xrightarrow{+2} 4 \xrightarrow{+3} 7 \xrightarrow{+4} 11$$

Therefore the next three terms are
$7 + 4 = 11, 11 + 5 = 16$ and $16 + 6 = 22$.
So the sequence is 1,2,4,7,11,16,22...

(d) The rule is alternating thus; subtract 3 from the first term, subtract 2 from the second term, subtract 3 from the third term, subtract 2 from the fourth term, and so on.

$$62 \xrightarrow{-2} 59 \xrightarrow{-3} 57 \xrightarrow{-2} 54 \xrightarrow{-3} 52$$

Therefore the next 3 terms are
$52 - 3 = 49, 49 - 2 = 47$, and $47 - 3 = 44$
So the sequence is 62, 59, 57, 54, 52, 49, 47, 44...

 Exercise 2:3

1. Determine the rule and use it to write down the next three terms in each of the following sequences.
   (a) 2,4,6,8 ...
   (b) 3,9,15,21, ...
   (c) 1,4,9,16 ...
   (d) 2,5,8,11,14,17, ...
   (e) 160,80,40,20 ...
   (f) 1,2,4,7, ...
   (g) 81,27,9,3 ...
   (h) 1,4,7,10, ...
   (i) 2,12,22,32, ...
   (j) 41,37,33,29 ...
   (k) 3,13,10,20 ...
   (l) 0, −4, −8, −12 ...
   (m) 3,5,9,15,23 ...
   (n) −1,3,7,11,15,..
   (o) 1, 3, 6, 10, 15,21, ...

2. The following pattern shows the maximum number of segments that can be drawn to connect a given number of points, no three of which lie on the same line. Without drawing the pattern, state the number of line segments that can be drawn to connect 6, 7 and 8 points.

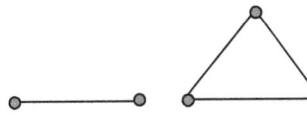 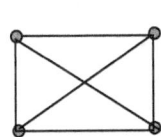 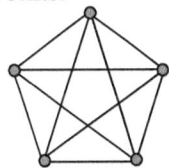

Multiple Choice Exercise 2

21. $\sqrt{7744}$ in index form is:
    [A] 26 [B] $2^3 \times 13$ [C] $2^3 \times 9$ [D] $2^3 \times 11$
22. $\sqrt{3136}$ in index form is:
    [A] 2×7^2 [B] $2^3 \times 7$ [C] 22×7 [D] 2×7
23. The square root $2\frac{7}{9}$ of is:
    [A] $1\frac{2}{9}$ [B] $1\frac{2}{3}$ [C] $1\frac{2}{5}$ [D] $1\frac{2}{7}$
24. The value of $\sqrt{5\frac{4}{9}}$ is:
    [A] $2\frac{1}{3}$ [B] $\frac{13}{18}$ [C] $5\frac{2}{5}$ [D] $16\frac{1}{3}$
25. The square root of $12\frac{1}{4}$ is:
    [A] $6\frac{1}{6}$ [B] $6\frac{1}{4}$ [C] $3\frac{3}{4}$ [D] $3\frac{1}{2}$
26. The square root of 0.0036 is:
    [A] 0.6 [B] 0.06 [C] 0.006 [D] 0.0006
27. The next 2 terms in the sequence 1, 2, 4, 7, 11, 16… are:
    [A] 17, 29 [B] 29, 24 [C] 22, 29 [D] 29, 40
28. The next term in the sequence 1, 4, 9, 16,…is:
    [A] 20 [B] 25 [C] 23 [D] 27
29. The next term in the sequence 2, 5, 11, 23, 47…is:
    [A] 95 [B] 93 [C] 71 [D] 27
30. In the sequence 1,3,7,15,31 the number that must be added to 31 to give the next term is:
    [A] 4 [B] 8 [C] 16 [D] 32
31. The number represented by * in the sequence 14, – 3, *, – 37 is:
    [A] 11 [B] –14 [C] 17 [D] – 20
32. The next three terms and the rule that describe the sequence 10, 20, 40, 80 . . . . are:
    [A] 82, 84, 86; start with 10 and add 2 repeatedly.
    [B] 90, 100, 110; start with 10 and add 10 repeatedly.
    [C] 320, 1280, 5120; start with 10 and multiply by 4 repeatedly.
    [D] 160, 320, 640; start with 10 and multiply by 2 repeatedly.
33. The next two terms of the sequence 1,5,14,30,55,... are respectively:
    [A] 61,110 [B] 67,116 [C] 81,140 [D] 91,140
34. The next three terms in the sequence 3,12,21,30,… are:
    [A] 40,50,60 [B] 38,46,54 [C] 39,48,57 [D] 36,32,39
35. The next four terms in the sequence −4,−1,2,5… are:
    [A] 5,8,11,14 [B] 8,11,14,17 [C] 3,6,9,12 [D] 0,8,11,14
36. If the pattern continues, the number of squares in the 8th diagram is:

Module 5, Topic 2: Number Patterns

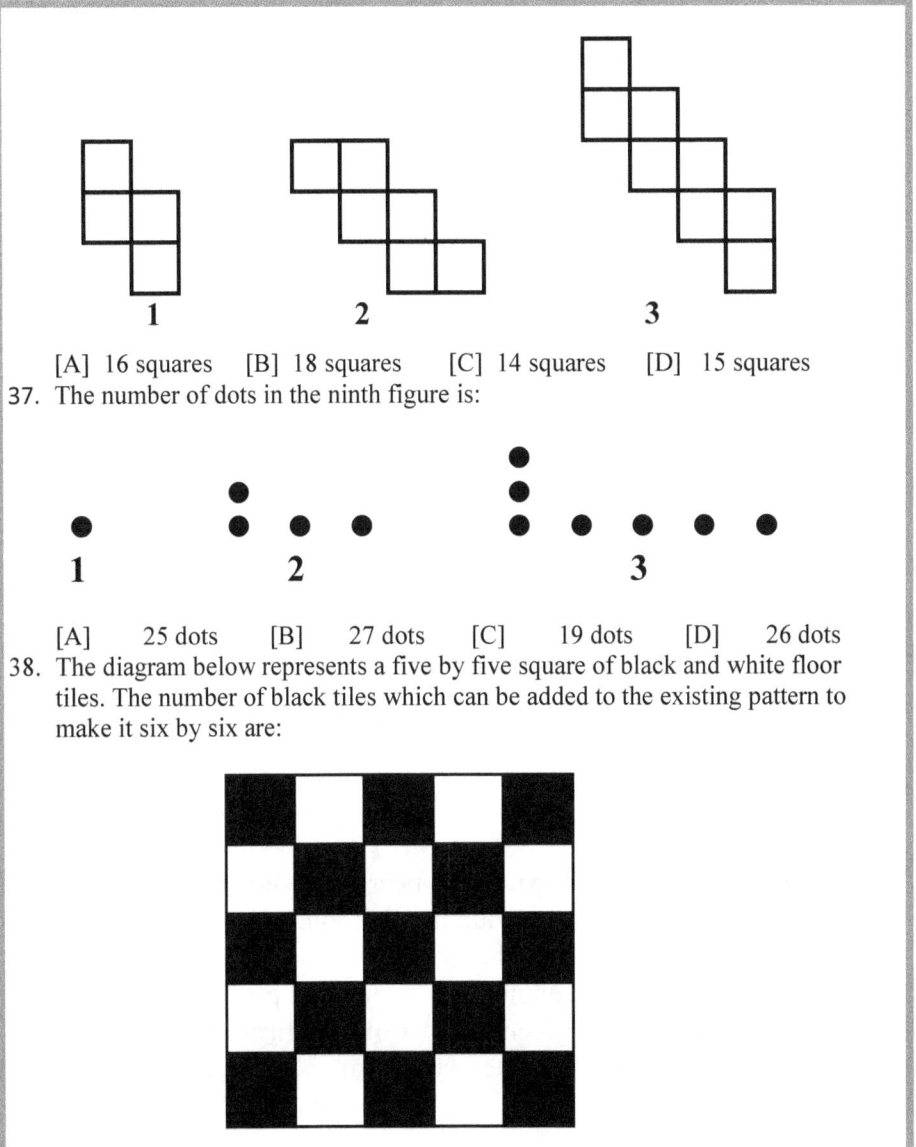

[A] 16 squares [B] 18 squares [C] 14 squares [D] 15 squares

37. The number of dots in the ninth figure is:

[A] 25 dots [B] 27 dots [C] 19 dots [D] 26 dots

38. The diagram below represents a five by five square of black and white floor tiles. The number of black tiles which can be added to the existing pattern to make it six by six are:

[A] 8 black tiles [B] 6 black tiles [C] 5 black tiles [D] 4 black tiles

Topic 3

FRACTIONS AND DECIMALS

Objectives

At the end of this topic, the learner should be able to:

1. Round-up or round-down to the nearest whole number, ten, hundred, thousand, tenth, hundredth, thousand etc.
2. Estimate the result of a calculation.
3. Approximate to a given number of decimal places.
4. Write a number to a given number of significant figures.
5. Write a given number in standard form.
6. Apply the knowledge of fractions and decimals to real life.

Module 5, Topic 3: Fractions and Decimals

 Review Exercise

1. Find the missing number in $\frac{2}{3} = \frac{}{15}$.

2. Simplify $\frac{28}{36}$.

3. Convert $5\frac{7}{8}$ to an improper fraction.

4. Compare the following using <, > or =.
   (a) $\frac{5}{8}$ and $\frac{3}{8}$ (b) $\frac{1}{6}$ and $\frac{3}{8}$ (c) $3\frac{7}{11}$ and $\frac{40}{11}$ (d) $5\frac{3}{4}$ and $\frac{71}{12}$

5. Evaluate the following.
   (a) $\frac{3}{5} + \frac{1}{5}$ (b) $\frac{9}{10} - \frac{2}{10}$ (c) $\frac{2}{3} + \frac{3}{5} + \frac{1}{6}$ (d) $\frac{7}{8} - \frac{3}{12}$
   (e) $\frac{5}{6} - \frac{2}{3} - \frac{1}{10}$ (f) $1\frac{3}{10} + 2\frac{1}{10}$ (g) $2\frac{2}{5} + 3\frac{1}{3}$ (h) $3\frac{2}{3} - 2\frac{3}{4}$ (i) $6\frac{1}{3} + 2\frac{1}{6} + 2\frac{2}{7}$ (j) $4\frac{3}{5} - 2\frac{1}{2} + 1\frac{1}{10}$

6. Evaluate $\frac{18}{35} \times \frac{14}{27}$.

7. Manka took $\frac{2}{3}$ of her money to school. At school, she used $\frac{3}{4}$ of what she took. What fraction of her money did she use?

8. Evaluate $5\frac{3}{8} \times 4$.

9. Mrs. Mundi had 367,000 francs in the credit union. She signed out $\frac{2}{3}$ of it. How much did she sign out?

10. Evaluate the following.
    (a) $4\frac{2}{3} \times 2\frac{3}{5}$ (b) $\frac{9}{10} + \frac{3}{5}$ (c) $1\frac{4}{5} + 3$

11. Out of the 13464 candidates who sat for an examination, 7854 passed. What fraction of the candidates failed?

12. Rewrite $\frac{347}{100}$ using a decimal marker.

13. What is the value of 4 in the number 6.75487?

14. Convert $2\frac{1}{25}$ to a decimal.

15. Change 0.125 to a fraction.

16. Compute the following without using a calculator.
    (a) 6.4163 + 7.3187 + 5.4128   (b) 8.9 − 2.47

17. Evaluate the following.
    (a) 136.8 × 47   (b) 1000 × 2.581   (c) 0.236 × 0.3

18. Find (a) $6\overline{)17.022}$ (b) $0.00025\overline{)0.4}$

19. Express the following numbers in standard form.
    (a) 10200   (b) 700000   (c) 0.0032   (d) 0.000073

3.1 Rounding Down and Rounding Up

? Brainstorming Exercise

The following diagram shows a number scale. Use the scale to answer the questions that follow.

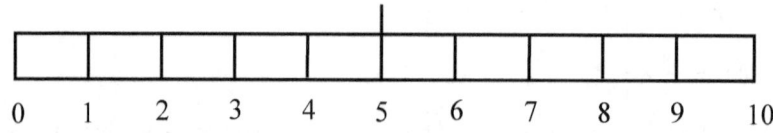

1. Which numbers are at the extreme ends of the scale?
2. Which number is in the middle of the scale?
3. List the numbers which are closer to 0 on the number scale.
4. List the numbers which are closer to 10 on the number scale.

For convenience and ease of calculations, numbers are usually **rounded down** by giving them smaller less exact values or **rounded up** by giving them larger less exact values. If the numbers are to be rounded to the nearest ten, then the numbers 1, 2, 3 and 4 which are closer to zero are written as 0 and the numbers 6, 7, 8 and 9 which are closer to ten are written as 10. Though 5, is midway, 5 is conventionally written as 10.

Approximation is the act of considering a number as its closest more convenient number, by means of rounding down or rounding up the number.

When the digit in the place after that required for the given degree of accuracy is any of 5, 6, 7, 8 or 9, we round up the digit required for the degree of accuracy by increasing it by 1 and ignore the rest in the case of decimals or replace them with zeros in the case of whole numbers.

When the digit in the place after that required for the given degree of accuracy is 0, 1, 2, 3 or 4, we round down the digit required for the degree of accuracy by maintaining this digit and ignoring the rest in the case of decimals or replace them with zeros in the case of whole numbers.

To round to the nearest ten, hundred, thousand…, replace the number by the closest multiple of 10, 100, 1000..., respectively.

To round to the nearest tenth, hundredth, thousandth, etc. write the number to 1, 2, 3, etc. decimal places.

3.2 Estimations and Approximations

An estimate is an intelligent guess on the dimensions of an object or the result of a calculation. When an estimate is made only an approximate (not exact) value is obtained.

To estimate the result of a calculation, round up or round down each number in the calculation to a reasonable or given degree of accuracy.

 Example

(a) *Estimating sums and differences in whole numbers*

1. Estimate the following
   (i) 245 + 350 + 570   (ii) 431 + 53   (iii) 748 – 394

   Solution
   (i) 245 + 350 + 570 ≈ 200 + 400 + 600 ≈ 1200
   (ii) 431 + 53 ≈ 430 + 50 ≈ 480
   (iii) 748 – 394 ≈ 700 – 400 ≈ 300

(b) *Estimating sums and differences in Decimals*

2. Estimate to the nearest whole number.
   (i) 4.68 + 0.71   (ii) 6.7234 – 3.5138

   Solution
   (i)   4.68 ≈ 5   (ii)   6.7234 ≈ 7
       + 0.71 ≈ +1   – 3.5138 ≈ –4
       ─────   ─────
          6   3
       ═════   ═════

3. Estimate to the nearest tenth.
   (i) 3.623 + 0.29 + 5.386   (ii) 4.86 – 3.456

   Solution
   (i) 3.623 + 0.29 + 5.386 ≈ 3.6 + 0.3 + 5.4 = 9.3
   (ii) 4.86 – 3.456 ≈ 4.9 – 3.5 = 1.4

Skill Building Exercise 3:1

1. Estimate the following sums and differences
   (a) 844 + 239   (b) 464 + 37   (c) 7982 + 1486
   (d) 5391 − 247   (e) 26212 − 14084   (f) 748 − 394
   (g) 509 − 42   (h) 13486 + 4842 + 29072
2. Estimate to the nearest whole number.
   (a) 0.89 + 2.083   (b) 6.4139 + 2.8238 + 3.2500
   (c) 4.9 − 0.87   (d) 8.7234 − 2.6006
3. Estimate to the nearest tenth.
   (a) 0.835   (b) 7.428   (c) 0.82   (d) 12.34
       + 0.27   6.38   −0.243   − 8.45
                    + 0.843

Example

(a) Estimating products and quotients In whole numbers

1. Estimate the following products
   (i) 58 × 24   (ii) 653 × 56   (iii) 48830 × 752   (iv) 275 × 24

   Solution
   (i) 58 × 24 ≈ 60 × 20 ≈ 1200   (ii) 653 × 56 ≈ 700 × 60 ≈ 42000
   (iii) 48,830 × 752 ≈ 50,000 × 800 ≈ 4,000,000
   (iv) 275 × 24 ≈ 300 × 20 ≈ 6,000

2. Estimate the following quotients
   (i) $\frac{562}{18}$ (ii) $\frac{68}{27}$ (iii) $\frac{62}{24}$

   Solution
   (i) $\frac{562}{18} \approx \frac{600}{20} \approx 30$ (ii) $\frac{68}{27} \approx \frac{70}{30} \approx 2.3$ (iii) $\frac{62}{24} \approx \frac{60}{20} \approx 3$

(b) Estimating of Products and Quotients in Decimals

1. Give an estimate for the following. (i) 4.755 × 0.5   (ii) $\frac{89.93}{4.1}$

   Solution
   (i) 4.755 × 0.5 ≈ 5 × 0.5 ≈ 2.5   (ii) $\frac{89.93}{4.1} \approx \frac{899.3}{41} \approx \frac{900}{40} \approx 22.5$

Module 5, Topic 3: Fractions and Decimals

 Skill Building Exercise 3:2

1. Estimate the following products.
   (a) 21 × 45   (b) 763 × 53   (c) 4.8 × 6
   (d) 0.21 × 3.81 (e) 8.6 × 5.9   (f) 3.1 × 8.2
2. Estimate the following quotients.
   (a) $32\overline{)85}$ (b) $91\overline{)873}$ (c) $3\overline{)6.31}$
   (d) $4.2\overline{)0.439}$ (e) $0.31\overline{)6.125}$ (f) $5.7\overline{)0.483}$

3.3 Decimal Places

The terms tenth, hundredth, thousandth etc. respectively mean one, two, three etc. decimal places. Thus rounding off to the nearest tenth, hundredth, thousandth etc. is the same as rounding off to 1, 2, 3 etc. decimal places.

3.4 Significant Figures

The significant figures of a given number are the figures from the leftmost non-zero digit to the rightmost non-zero digit (or rightmost zero digit in the case where the digit on the required degree of accuracy is zero) required to express the number to a given degree of accuracy. This definition implies that, 4.071 to four significant figures for instance, is the same as 4.0710 to five significant figures and 4068 to three significant figures for instance is the same as 4070 to four significant figures.

A zero between numbers is a significant figure e.g. the zero in 4068 is significant. On the other hand a zero at the end of a number is not significant. Thus 6530 is written to 3 significant figures.

 Example

1. Round 6526 to the nearest:
   (a) ten   (b) hundred   (c) thousand.
   In each case, state the number of significant figures.

   Solution
   (a) 6526 to the nearest ten = 6530   [3 significant figures]
   (b) 6526 to the nearest hundred = 6500   [2 significant figures]
   (c) 6526 to the nearest thousand = 7000   [1 significant figure]

2. Round 1.045 to:
   (a) 2 decimal places (b) 1 decimal place
   In each case, state the number of significant figures.

   Solution
   (a) 1.045 to 2 decimal places = 1.05   [3 significant figures]
   (b) 1.045 to 1 decimal place = 1.0   [1 significant figure]

3. Round 0.01027 to:
   (a) 4 decimal places.   (b) 3 decimal places.
   (c) 2 decimal places.   (d) 1 decimal place.
   In each case, state the number of significant figures.

   Solution
   (a) 0.01027 to 4 decimal places = 0.0103   [4 significant figures]
   (b) 0.01027 to 3 decimal places. = 0.010   [2 significant figure]
   (c) 0.01027 to 2 decimal places = 0.01   [2 significant figures]
   (d) 0.01027 to 1 decimal place = 0.0   [no significant figures]

4. Round 0.2007043 to:
   (a) 6 significant figures.   (b) 5 significant figures.
   (c) 4 significant figures.   (d) 3 significant figures.
   (e) 2 significant figures.   (f) 1 significant figure.

   Solution
   (a) 0.2007043 = 0.200704 to 6 significant figures
   (b) 0.2007043 = 0.20070 to 5 significant figures
   (c) 0.2007043 = 0.2007 to 4 significant figures
   (d) 0.2007043 = 0.201 to 3 significant figures
   (e) 0.2007043 = 0.20 to 2 significant figures
   (f) 0.2007043 = 0.2 to 1 significant figure

Notice that (b) and (c) are virtually the same and (e) and (f) are virtually the same.

Module 5, Topic 3: Fractions and Decimals

 Skill Building Exercise 3:3

1. Complete the following table.

Number		Number of significant figures			
		1	2	3	4
a	0.0068398				
b	2.0068398				
c	4.69768				
d	1.006127				

2. Complete the following table.

Number		Number of Decimal places			
		1	2	3	4
a	0.0068398				
b	2.0068398				
c	4.69768				
d	1.006127				

3. Express to 3 decimal places.
   (a) 0.003646   (b) 0.4567   (c) 0.5046
4. Express to 2 decimal places.
   (a) 14.9028   (b) 23.1058   (c) 6.0381
5. Express to 3 significant figures.
   (a) 0.02485   (b) 4.027956
6. Express to 2 significant figures.
   (a) 547.53   (b) 59.81798   (c) 5382
7. Express to 1 significant figure.
   (a) 0.009238   (b) 5.097   (c) 0.2309

3.5 Standard Form

In section 6.6 of book 1 we saw that, to express a number in standard form or scientific notation,

(i) Place the decimal point immediately after the first non-zero digit and write down the product
Number from the first non-zero digit to the last non-zero digit × 10

(ii) From immediately after the first non-zero digit in the number, count the number of places up to the decimal point or to where the decimal point is supposed to be in the cases of whole numbers and raise 10 to this power placing a negative (−) sign in front of the power if you counted to the left or don't place any sign if you counted to the right.

 Skill Building Exercise 3:4

Express the following numbers in standard form.
(a) 5000   (b) 480   (c) 10200   (d) 700000
(e) 0.0032   (f) 0.000073   (g) 0.925   (h) 0.00011
(i) 0.5600   (j) 3000×10^{-8} (k) 19.6×10^{-4} (l) 0.034×10^{-2}

Calculations in standard form

In performing calculations in standard form the powers of 10 are manipulated using the multiplication and division laws of indices.

 Example

Evaluate the following leaving your answers in index form.
(a) $10^3 \times 10^5$ (b) $10^{-6} \times 10^4$ (c) $10^{-8} \times 10^{-3}$
(d) $10^3 \div 10^5$ (e) $10^{-6} \div 10^4$ (f) $10^{-8} \div 10^{-3}$

Solution
(a) $10^3 \times 10^5 = 10^{3+5} = 10^8$ (b) $10^{-6} \times 10^4 = 10^{-6+4} = 10^{-2}$
(c) $10^{-8} \times 10^{-3} = 10^{-8+(-3)} = 10^{-11}$ (d) $10^3 \div 10^5 = 10^{3-5} = 10^{-2}$
(e) $10^{-6} \div 10^4 = 10^{-6-4} = 10^{-10}$ (f) $10^{-8} \div 10^{-3} = 10^{-8-(-3)} = 10^{-5}$

 Skill Building Exercise 3:5

1. Evaluate giving your result in standard form.

(a) $9.5 \times 10^7 - 3.08 \times 10^6$ (b) $\dfrac{0.45 \times 0.91}{0.0117}$ (c) 0.06×0.09

(d) $\dfrac{0.24}{0.012}$ (e) $\dfrac{8.75}{0.025}$ (f) $\sqrt{\dfrac{0.81 \times 10^5}{2.25 \times 10^7}}$

(g) $\dfrac{9.6 \times 10^8}{0.24 \times 10^5}$ (h) $\dfrac{0.9687}{0.001}$ (i) $\dfrac{0.203 \times 0.55}{3.05}$

(j) $\sqrt{\dfrac{1.44 \times 10^5}{8.1 \times 10^4}}$ (k) $\sqrt{\dfrac{0.0016 \times 0.0081}{0.36}}$ (l) $\sqrt{\dfrac{76.42 \times 10^{-1}}{0.004 \times 10^2}}$

Module 5, Topic 3: Fractions and Decimals

2. Find the value of $\dfrac{6\times10^3\times2\times10^{-4}}{4\times10^{-5}}$ expressing your answer in standard form.
3. The weight of an atom of hydrogen is about 0.00000000000000000000000017 grams. Express this weight in standard form.
4. Evaluate $\dfrac{1946\times10^{-1}}{2\times10^2}$, giving the answer
   (a) In standard form.   (b) Correct to 1 decimal place.
   (c) Correct to 2 significant figures.
5. Evaluate $\dfrac{12.78\times10^{-3}}{9\times10^{-1}}$, expressing your answer
   (a) In standard form   (b) Correct to two significant figures
   (c) Correct to three decimal places

3.6 Application of Fractions and Decimals to Real Life

 Integration Activity

1. Awa and sons and New Life Enterprise both sell the same type of a food flask. At Awa and sons the food flask is marked 16000 francs, 25% discount. At New Life Enterprise the food flask is marked 18300 francs, $\frac{1}{3}$ discount. From which shop would you prefer to buy the flask?
2. Dreamland restaurant which normally sells a dish of food at 1250 francs decides to give an offer of two dishes at the same price to its customers. Alizan restaurant which normally sells a dish of food at 1150 francs decides to give an offer of 50% off the price to its customers. You and two of your friends want to go to a restaurant. Which of the two restaurants would you chose?
3. The marks in the report card of a student are as follows. Compare his results in the three subjects.

Subject	Mark
Chemistry	$\dfrac{28}{36}$
Geography	$\dfrac{34}{40}$
Literature	72%

4. At the beginning of the school year a shop keeper buys 80 school bags at a total cost of 160000 francs. He sells $\dfrac{2}{5}$ of them for 4000 francs each, 32% of them for

3000 francs each and the rest for 2500 francs. What is his profit on the sales of these bags?

5. A man draws the plan of a snack bar as follows. He intends that the private room should be $\frac{3}{8}$ of the whole plan. What must the length of the main store be?

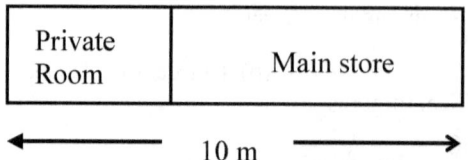

6. A bush faller hires a car with a full tank of fuel. The conditions are that he will bring back the car with the tank $\frac{1}{4}$ full of fuel. The bush faller uses nearly all the petrol in the tank. He estimates that a full tank is about 45 litres of petrol. Approximately how many litres of fuel should the bush faller put into the tank?

7. Statistics show that approximately 27% of a certain community smoke. The following data was published in a report published after a campaign against smoking youths by the Ministry of health.

	Before Campaigns	After Campaigns
Fraction of youths who smoke	$\frac{2}{5}$	$\frac{1}{4}$
Number of youths who smoke	4000	3200

(i) Compare the number of youths who smoked before the campaigns to the prevalence rate of the society.

(ii) Compare the number of youths who smoked after the campaigns to the prevalence rate of the society.

Module 5, Topic 3: Fractions and Decimals

 Multiple Choice Exercise 3

1. 0.015849 expressed correct to three significant figures is:
   [A] 0.0158   [B] 0.0159   [C] 0.0160   [D] 0.020
2. 6474 correct to three significant figures is:
   [A] 647   [B] 648   [C] 6470   [D] 6480
3. The number 25.973 correct to three significant figures is:
   [A] 25.973   [B] 25.97   [C] 25.9   [D] 26.0
4. The number of people attending a football match is 27000, correct to 2 significant figures. The greatest possible attendance shown by this figure is:
   [A] 27,000   [B] 27499   [C] 27599   [D] 26999
5. 0.0063 correct to 2 decimal places is:
   [A] 0.006   [B] 0.01   [C] 0.06   [D] 0.10
6. After evaluating 2.35 × 0.48, the answer to 2 decimal places is:
   [A] 11.28   [B] 1.13   [C] 1.128   [D] 1.10
7. The value of 3.769 ÷ 0.7 to the nearest tenth is:
   [A] 5.41   [B] 5.0   [C] 10   [D] 5.4
8. To the nearest whole number, the result of $\dfrac{6.6 \times 1.8}{5.4}$ is:
   [A] 2.2   [B] 3   [C] 2   [D] 22
9. 0.000252 ÷ 0.007 to two decimal places is:
   [A] 0.04   [B] 0.03   [C] 0.36   [D] 0.40
10. By evaluating $\dfrac{7 + 3.32}{9.91 - 5.11}$, the answer to one decimal place will be:
    [A] 21.5   [B] 2.1   [C] 22.0   [D] 2.2
11. 0.44734 ÷ 0.01, evaluated to the nearest hundredth is:
    [A] 44.70   [B] 45   [C] 44.73   [D] 44.00
12. $\dfrac{6.3 \times 60 \times 0.2}{3.6 \times 1.4}$, when simplified, the answer to the nearest ten is:
    [A] 15   [B] 20   [C] 10   [D] 1.5
13. 930,000,000 in standard form is:
    [A] 93.0×10^9 [B] 9.3×10^8 [C] 9.3×10^7 [D] 9.3×10^{-7}
14. 5238, expressed in standard form is:
    [A] 5.238×10^3 [B] 5.238×10^2 [C] 5.238×10^1 [D] 5.238×10^0
15. Expressed in standard form 435600 is:
    [A] 4.536×10^7 [B] 4.536×10^6 [C] 4.536×10^5 [D] 4.536×10^4
16. When expressed in standard form 2789 equals:
    [A] 2.789×10^{-3} [B] 2.789×10^2 [C] 2.789×10 [D] 2.789×10^3
17. In standard form, 52006 is equal to:
    [A] 5.2006×10^3 [B] 5.2006×10^{-4} [C] 5.2006×10^4 [D] 5.2006×10^{-3}
18. 120,000 written in standard form is:
    [A] 1.2×10^2 [B] 1.2×10^3 [C] 1.2×10^4 [D] 1.2×10^5
19. The number 36700 written in standard form is:

[A] 3.67×10^3 [B] 3.67×10^5 [C] 3.67×10^4 [D] 3.67×10^2

20. 325,000 in standard form is:

    [A] 3.25×10^6 [B] 3.25×10^{-6} [C] 3.25×10^5 [D] 3.25×10^{-5}

21. 0.00562 in standard form is:

    [A] 5.62×10^{-3} [B] 0.562×10^{-2} [C] 5.62×10^{-2} [D] 5.62×10^2

22. Expressed in standard form 0.0462 is:

    [A] 0.462×10^{-1} [B] 0.462×10^{-2} [C] 4.62×10^{-1} [D] 4.62×10^{-2}

23. 0.000834 in standard form is:

    [A] 8.34×10^{-4} [B] 8.34×10^{-5} [C] 8.34×10^3 [D] 8.34×10^4

24. 0.0000027 in standard form is:

    [A] 2.7×10^6 [B] 2.7×10^{-6} [C] 2.7×10^5 [D] 2.7×10^{-5}

25. 0.000,000,070,2 in standard form is:

    [A] 7.02×10^{-5} [B] 7.5×10^{-6} [C] 7.02×10^{-7} [D] 7.5×10^{-8}

26. Expressed in standard form 0.000,082,3 becomes:

    [A] 0.823×10^5 [B] 0.823×10^{-5} [C] 8.23×10^5 [D] 8.23×10^{-5}

27. Written in standard form 0.000370 is:

    [A] 3.7×10^{-1} [B] 7.5×10^{-2} [C] 3.7×10^{-3} [D] 7.5×10^{-4}

28. 46 × 900 expressed in standard form is:

    [A] 4.14×10^3 [B] 4.14×10^5 [C] 4.14×10^4 [D] 4.14×10^6

29. In standard form, converting 4 hours to seconds the result is:

    [A] 1.44×10^3 [B] 1.44×10^{-3} [C] 1.44×10^4 [D] 1.44×10^{-4}

30. 258 km when expressed to mm and expressed in standard form becomes:

    [A] 2.58×10^8 [B] 2.58×10^7 [C] 2.58×10^6 [D] 2.58×10^5

31. $\dfrac{8.75}{0.025}$, expressed in standard form is:

    [A] 3.5×10^{-3} [B] 3.5×10^{-2} [C] 3.5×10^1 [D] 3.5×10^2

32. The product of 0.06 and 0.09 in standard form is:

    [A] 5.4×10^2 [B] 5.4×10^{-3} [C] 5.4×10^1 [D] 2.65×10^4

33. Evaluating $0.009 \div 0.012$, the answer in standard form is:

    [A] 7.5×10^2 [B] 7.5×10^{-3} [C] 7.5×10^{-1} [D] 7.5×10^{-2}

34. 5.7×10^4 in ordinary form is:
    [A] 5700   [B] 57000   [C] 7500   [D] 75000

35. As a decimal fraction 8.2×10^{-5} is:
    [A] 0.0082   [B] 0.00082   [C] 0.000082   [D] 0.0000082

36. In normal form 3.746×10^{-3} is:
    [A] 0.003746   [B] 0.0003746   [C] 0.03746   [D] 3746

37. 9.258×10^{-3} in normal form correct to 3 significant figures is:
    [A] 926   [B] 0.093   [C] 0.009   [D] 0.00926

38. 7.15×10^5 in normal form is:

Module 5, Topic 3: Fractions and Decimals

    [A] 71500   [B] 715000   [C] 7150   [D] 7150000

39. The number 4.5×10^3 is the same as:
    [A] 0.0045   [B] 0.045   [C] 4500   [D] 45000

40. The sum of 728.93 and 0.46 expressed in standard form to three significant figures is:
    [A] 72.9×10 [B] 7.29×10^2 [C] 728×10^0 [D] 7.28×10^0

41. The sum of 2.48×10^3 and 5.9×10^4 is:
    [A] 6.148×10^1 [B] 6.148×10^4 [C] 6.148×10^3 [D] 6.148×10^2

42. $4 \times 10^2 \times 2 \times 10^{-4}$ is equal to:
    [A] 8×10^6 [B] 8×10^{-2} [C] 8×10^{-8} [D] 6×10^{-2}

43. $7.580 \times 10^9 + 7.677 \times 10^9$ is equal to:
    [A] 1.5257×10^{10} [B] 1.5257×10^8 [C] 1.5257×10^9 [D] 1.5257×10^7

44. $5.7 \times 10^6 - 1.8 \times 10^6 = :$
    [A] 3.9×10^{-6} [B] 3900000×10^6 [C] 3.9×10^6 [D] 39×10^5

45. 252000 is equal to:
    [A] 2.52×10^5 [B] 2.52×10^6 [C] 2.52×10^4 [D] 2.52×10^{-5}

46. The product of 0.012 and 0.0008 in standard form is:
    [A] 9.6×10^6 [B] 9.6×10^{-6} [C] 9.6×10^5 [D] 9.6×10^{-5}

47. 450×70 is:
    [A] 3.15×10^5 [B] 3.15×10^4 [C] 3.15×10^3 [D] 3.15×10^2

48. The result of $\dfrac{0.126}{36}$ is:
    [A] 3.5×10^4 [B] 3.5×10^{-4} [C] 3.5×10^3 [D] 3.5×10^{-3}

49. A man spends $\dfrac{3}{4}$ of his monthly salary on food and $\dfrac{1}{2}$ of the remainder on rent. If he has 15000 FRS left, the amount he earns is:
    [A] 60,000 FRS   [B] 90,000 FRS   [C] 105,000 FRS   [D] 120,000 FRS

50. Given that a pole has $\dfrac{1}{3}$ of its length in mud $\dfrac{2}{5}$ of the remainder in water and the rest 6 m long above the surface of the water, the length of the pole is:
    [A] 12 m   [B] 10 m   [C] 15 m   [D] 16 m

Topic 4

ARITHMETIC PROCESSES

Objectives

At the end of this topic, the learner should be able to:

1. Solve real life problems involving direct and inverse proportion.
2. Calculate profit or loss after selling an article
3. Define the terms principal, interest, time, rate and amount.
4. Identify the principal, interest, time, rate and amount in problems.
5. Calculate the simple interest on a given amount for a given period at a given rate.
6. Solve real life problems involving simple interest.
7. Calculate the compound interest on a given amount for a given period at a given rate.

Module 5, Topic 4: Arithmetic Processes

4.1 Variation in Real Life

> **? Brainstorming Exercise**
>
> In real life, there are many quantities which relate, in one way or the other.
> 1. Name as many pair of quantities as possible which are related in such a way that one increases as the other increases.
> 2. Explain how these quantities are related.
> 3. Name as many pair of quantities as possible which are related in such a way that one increases as the other decreases.
> 4. Explain how these quantities are related.

In real life, many quantities are related to each other. Examples of some related quantities in real life are:

The interest generated per year increases as the amount of money invested increases.

The time taken to complete a given piece of work reduces, as the number of people doing the work increase.

The electric bill increases as the amount of current used increases.

The area of the circle increases as the radius increases.

Temperature decreases with increasing altitude (i.e. the higher you go, the colder it becomes)

The amount of current passing in a wire increases with decreases in resistance.

The above are some examples of variations. Many other examples of variation exist in natural science. You may be interested to have an idea of a few of the natural science laws that deal directly with variation. Some of them include,

Hooke's law states that the length of a spring or elastic increases as the force applied to it increases.

Charles's law: states that the volume of a fixed number of particles of gas is directly proportional to the absolute temperature.

Boyle's law states that the pressure of a fixed mass of gas at constant temperature increases as the volume decreases, and vice versa.

There are two basic types of variation.

- Direct variation or direct proportion and
- Inverse variation or inverse proportion

4.2 Direct Variation

 Investigative Activity

1. One pen cost 100 francs. Complete the following table to show the prices of 1,2,3 ... 10 pens.

Number of pens, n	Cost of pens in francs
1	100
2	
3	
4	
5	
6	
7	
8	
9	
10	

2. What conclusion do you draw concerning the price as the number of pens increase?

From the above investigation, we see that the number of pens increases as the price increases. This is an example of **direct variation** or **direct proportion**. Notice that the changes in each case are related by a constant multiplier (in this case 100). This constant multiplier is called the **constant of proportionality** or **coefficient of proportionality**.

4.3 Using the Unitary Method to Solve Proportions

The unitary method involves finding the proportion of a given quantity by first finding the value of one part of the whole called the **unit**. In this method the quantity we want to find is always written at the end of each statement. In simple proportions, three statements suffice to solve the problem. The second statement which seeks to find the value per unit is called the unitary statement.

Module 5, Topic 4: Arithmetic Processes

 Example

A woman is paid 20,000 **FRS** for two days of work. How much will she be paid if she works for 10 days?
Solution
Amount paid for 2 days = 20,000 FRS.
Amount paid for 1 day = $\frac{20,000}{2}$ FRS.   [Unitary statement]
Therefore, amount paid for 10 days = $\frac{20,000}{2} \times 10 = 100,000$ FRS.

 Exercise 4:1

1. The cost of 5 packets of sugar is 4500 FRS. What will be the cost of 11 packets?
2. A labourer is paid 22500 FRS for 9 days of work. For how many days should he work to be paid 42500 FRS.?
3. 11 books cost 3300 FRS. What will be the cost of 36 books?
4. Mr. Ngwa earns 2,000 FRS. For 4 hours of work. For how long, will he work to earn;   (a) 10,000 FRS.   (b) 20,000 FRS.   (c) 15,000 FRS.
5. A car travels 80km in 2 hours. How far will it travel in;
   (a) 1 hour   (b) 3 hours   (c) 12 hours
6. Three mathematics books cost 600 Frs. What is the cost of;   (a) 6 books   (b) 10 books   (c) 12 books
7. Four teachers can parcel 5600 question papers in a day.
   (a) How many questions papers can 7 teachers working at the same rate parcel in a day?
   (b) How many teachers working at the same rate will be required to parcel 23,800 question papers in one day?

Real life Situations

1. A bank wanted to give loans of 1,000,000 FCFA and 600,000 FCFA to Mr. Fombe and Mr. Tarla respectively. Later due to a saving of 1,120,000 FCFA with the bank, the bank manager decides to use all this money to increase the loans of Mr. Fombe and Mr. Tarla. Given that the ratio of the loans must be maintained, calculate the total amount the bank will give to each of them.
2. A pig farmer uses 200 kg of food to feed 450 pigs.
   (a) How many pigs, is he expected to feed with 328 kg of feed?
   (b) How many kilograms of feed, is he expected to feed 630 chicks with?

4.4 Inverse Variation

 Investigative Activity

Given that one person can take 60 minutes to fill a tank with water.
2. Complete the following table to show the amount of time (t) in minutes that 1, 2, 3, 4, 5 and 6 people all working at the same rate can use to fill a similar tank.

Number of People, P	1	2	3	4	5	6
Time taken, t (mins)	60					

3. What conclusion do you draw concerning the amount of time spent as the number of people increase?

From the investigation above we can see that the time decreases as the number of people increases. This is an example of **inverse variation** or **inverse proportion**. In this case coefficient of proportionality is 600.

 Example

A bag of rice feeds 40 people for 2 weeks. How long will it take to feed 8 people?

Solution
40 people eat a bag of rice for 2 weeks.
1 person eats a bag of rice for 2×40 days
8 people will eat a bag of rice for $\frac{2 \times 40}{8} = 10$ weeks.

Exercise 4:2

1. 4 men dig a well in 5 days. How long will it take
   (a) 1 man   (b) 10 men   (c) 20 men, to dig the well.
2. A boy takes 15 minutes to walk to school at a speed of 6 km/hr. how long will it tale if he uses a bicycle at a speed of
   (a) 15 km/hr.   (b) 30 km/hr.   (c) 45 km/hr.
3. 2 boys take 60mins to fill a drum of water. How many minutes will it take to fill the drum if there are   (a) 20 boys   (b) 60 boys   (c) 40 boys.
4. 10 people working at the same rate mould 3600 blocks in six days. How long would it take 15 people to mould the same amount?
5. 4 women hoe a farm in 30 hours. How many women working the same rate would hoe a similar farm in one day?
6. A basket of guavas is shared equally amongst 8 students and each receives 9 guavas. If the guavas were divided amongst 12 students, how many guavas would each student receive?
7. 20 workers of a soap factory produce 3000 tablets of soap in 12 hours. How long will it take 15 workers to produce the same number of tablets?

4.5 Profit and Loss

When a businessman buys goods for resale, the price at which the goods are bought is called the **cost price** while the price at which the goods are sold is called the **selling price**. Though his objective is to sell the goods at a selling price, which is higher than the cost price, he might end up selling them at a selling price, which is less than the cost price.

If his selling price is more than the cost price, he is said to have made a **profit** or **gain**.
Thus,

$$\text{Profit} = \text{Selling Price} - \text{Cost Price}$$

If his selling price is less than the cost price, he is said to have made a **loss**.

$$\text{Loss} = \text{Cost Price} - \text{Selling Price}$$

Example

1. A dealer bought a car at 3,200,000 Frs. and sold it at 3,900,000 Frs. What was his profit?

   Solution
   Profit = selling price − cost price = 3,900,000 − 3,200,000 = 700,000 Frs.

2. A trader bought 15 bags of rice at 12,500 Frs. each. If she sold all these bags for 180,000 Frs., what profit or loss, did she make?

   Solution
   Total cost price for 15 bags = 12,500×15=187,500 Frs.
   Total selling price for 15 bags = 180,000 Frs.
   Cost price is greater than the selling price so she made a loss.
   Loss = cost price − selling price = 187,500 − 180,000 = 7,500 Frs.

4.6 Percentage Profit And Loss

The percentage profit or loss is the ratio of the profit or loss to the cost price expressed as a percentage.

Let the cost price be CP and the selling price be SP.

$$\text{Percentage profit} = \frac{\text{profit}}{\text{CP}} \times 100\% = \frac{\text{SP} - \text{CP}}{\text{CP}} \times 100\%$$

$$\text{Percentage loss} = \frac{\text{loss}}{\text{CP}} \times 100\% = \frac{\text{CP} - \text{SP}}{\text{CP}} \times 100\%$$

It follows that:

(1) A profit of $p\%$ means the SP is $(100 + p)\%$ of the CP. In this case,

$$SP = \frac{(100+p)}{100} \times CP \quad \text{and} \quad CP = \frac{100}{(100+p)} \times SP$$

(2) A loss of $l\%$ means the SP is $(100 - l)\%$ of the CP. In this case,

$$SP = \frac{(100-p)}{100} \times CP \quad \text{and} \quad CP = \frac{100}{(100-p)} \times SP$$

Example

1. A Woman bought a car at 4,500,000 Frs. and sold it at a profit of 15%. What was its selling price?

   Solution
   $$SP = \frac{100+p}{100} \times CP = \frac{100+15}{100} \times 4,5500,000 = 5,175,000 \text{ Frs.}$$

2. A man sold a house for 3,600,000 Frs. incurring a loss of 40 %. Calculate the amount he paid for the house.

   Solution
   $$CP = \frac{100}{(100-p)} \times SP = \frac{100}{60} \times 3600000 = 6,000,000 \text{ Frs.}$$

3. A trader bought a television at 250,000 Frs. and sold it for 280,000 Frs. Calculate his percentage profit or loss.
   Solution
   He made profit because his selling price is higher than the cost price.
   $$\text{Percentage profit} = \frac{SP-CP}{CP} \times 100\% = \frac{280,000-250,000}{250,000} \times 100\% = 12\%$$

 Exercise 4:3

1. In 2006, the population of a certain village was 12,500 people. In 2009, the population was 15,500 people. Find the percentage increase in the population of the village.
2. A businessman had 720 customers last year. Due to poor management, the number of customers reduced to 600 this year. Calculate the percentage decrease in the number of customers.
3. An article costing 24,000 FCFA depreciated to 17,000 FCFA at the end of the year. By what percentage did it depreciate?
4. To convert 4 km/h to m/s, a student obtained an answer 1 m/s. Calculate his percentage error.
5. After a slim course the weight of a woman who originally weighed 89 kg 600 g, reduced to 78 kg 400 g. Calculate the percentage decrease in her weight.
6. The enrolment of a school in 1995 was 18,525 students and in 2005, it was 22,750 students. Calculate the percentage increase in the population of the school for this period correct to 1 decimal place.
7. During a Physics practical, a student measured the length of a wire as 18 mm instead of 20 mm.
   What is his percentage error in the measurement of the wire?
8. Yaje bought a bag of huckleberry at 6,000 FCFA and retailed it for 5,800 FCFA. Calculate her percentage profit or loss.

4.7 Basic Terms Related to Interest

Principal (P)

Principal is the amount of money, which is invested (lent or borrowed).

Interest (I)

Interest is the charge by an investor or lender on his investment. The one who borrows the money incurs the interest.

Time (T)

Time is the duration or period for which an investment is made.

Rate (R)

Rate is the ratio of the interest to the principal, expressed as a percentage for a given period. Though the period is usually one year (per annum), it could be any other period say per month or per week. For instance, 12% per annum, 5% per month, etc.

Amount (A)

Amount is the sum of the principal and the interest after a given period.

$$A = I + P$$

4.8 Simple Interest

Simple interest is the interest charged on the principal for a given period. With simple interest, unless the investor increases the principal, the principal will remain the same no matter the number of years the investment (or loan) lasts. Interest increases with the principal, the time and the rate. Sometimes the rate is per annum while the time is in months or days. In such a case, it is necessary to convert the time to years by dividing by 12 or 365 as the case may be.

Thus,

$$I = \frac{PRT}{100} \Leftrightarrow P = \frac{100I}{RT}, T = \frac{100I}{PR}, R = \frac{100I}{PT}$$

Example

1. Calculate the simple interest, which Mrs. Fube will get on 640,000 FRS for 2 years 6 months at the rate of $4\frac{1}{2}\%$ per annum.

   Solution

   $P = 640,000$ Frs, $T = 2\frac{1}{2}$ years, $R = 4\frac{1}{2}\%$

   $I = \dfrac{PTR}{100} = \dfrac{640000 \times 2\frac{1}{2} \times 4\frac{1}{2}}{100} = \dfrac{640000}{100} \times \dfrac{5}{2} \times \dfrac{9}{2} = 72000$ Frs

2. Mrs. Danjuma paid a simple interest of 132, 000 FRS., for money borrowed at 8% per annum after 3 years. Calculate the sum of money she borrowed.

   Solution
   $I = 132,000$ Frs., $T = 3$ years,
   $R = 8\%$ per annum.

   $P = \dfrac{100I}{RT} = \dfrac{100 \times 132000}{3 \times 8} = 550,000$ Frs.

3. Mr. Jaiy invested 700,000 Frs. at 4% per annum simple interest. How long will the amount reach 784000 Frs.?

   Solution
   $A = 784,000$ Frs., $P = 784,000$ Frs., $R = 4\%$

   $I = A - P = 784,000 - 700,000 = 84,000$ Frs.
   $T = \dfrac{100I}{PR} = \dfrac{100 \times 84,000}{700,000 \times 4} = 3$ years

4. At what rate must Pa Fru invest 2,500,000 Frs. for 4 years to yield a simple interest of 500,000 Frs.?

   Solution
   $I = 500,000$ Frs., $P = 2,500,000$ Frs.,
   $T = 4$ years

   $R = \dfrac{100I}{PT} = \dfrac{100 \times 500,000}{2,500,000 \times 4} = 5\%$ per annum.

5. Mr. Nfor invested 700,000 Frs. in a Credit Union for 36 months at 6% per annum. Calculate his simple interest for this period

Solution

$T = 36 \text{ months} = \dfrac{36}{12} \text{ years} = 3 \text{ years}$

$P = 700{,}000 \text{ Frs}, \; R = 6\%$

$I = \dfrac{PRT}{100} = \dfrac{700{,}000 \times 6 \times 3}{100} = 126{,}000 \text{ Frs}.$

 Exercise 4:4

1. Find the simple interest on:
   (a) 300,000 FCFA at 12 % per annum for 5 years.
   (b) 160,000 FCFA at 8 % per annum for 9 months.
   (c) 400,000 FCFA at 10 % per annum for 18 months.
2. What principal will generate an interest of 15,000 FCFA for two months at 9 % per annum simple interest?
3. At what rate per annum can 73,000 FCFA yield an interest of 4800 FCFA for 240 days?
4. How long will 1,000,000 FCFA yield an interest of 40,000 FCFA at 8 % per annum?
5. In the following table, the interest is per annum. Solve for the missing item.

	P (FCFA)	R	T	I (FCFA)
a	500,000	?	6 months	30,000
b	?	7 %	1.5 years	21,000
c	5,000,000	10 %	?	1,000,000

4.9 Compound Interest

Compound interest is the money paid on the principal and the interest accumulated from the past years or months as the case may be. While the simple interest remains the same over the whole period, the compound interest increases from year to year as the amount increases.

Progressive Method for Compound Interest

To calculate compound interest using the progressive method, we calculate the principal at the beginning of each year and use it to compute the interest for the year. The compound interest is then the sum of all the interest for all the years.

 Example

Mr. Mofor invested the sum of 3,000,000 Frs. for 3 years at 10% per annum compound interest. Calculate his interest at the end of the 3 years.

Solution

Let the principals for first, second and third years be P_1, P_2, P_3 and interest for first, second and third years be I_1, I_2 and I_3 respectively.

Then,

$$R = 10\%, P = 3,000,000 \text{ Frs}, I = PRT = \frac{10}{100}P$$

$$P_1 = 3,000,000 \text{ Frs} \Rightarrow I_1 = \frac{10 \times 3,000,000}{100} = 300,000 \text{ Frs.}$$

$$P_2 = P_1 + I_1 = 3,000,000 + 300,000 = 3,300,000 \text{ Frs}$$

$$\Rightarrow I_2 = \frac{10 \times 3,300,000}{100} = 330,000 \text{ Frs.}$$

$$P_3 = P_2 + I_2 = 3,300,000 + 330,000 = 3,630,000 \text{ Frs}$$

$$\Rightarrow I_2 = \frac{10 \times 3,630,000}{100} = 363,000 \text{ Frs.}$$

Compound interest $= I_1 + I_2 + I_3$

$$= 300,000 + 330,000 + 363,000 = 993,000 \text{ Frs.}$$

Alternatively,
The amount after three years is
$A_3 = I_3 + P_3 = 3630000 + 363000 = 3993000$
Compound interest $I = A - P = 3,993,000 - 3,000,000 = 993,000$ Frs.

4.10 The Compound Interest Formula

Suppose that we invest an amount P at the rate of $r\%$ per annum. Then the formula for calculating the amount after t years is given by

$A = P\left(1 + \frac{r}{100}\right)^t$ and $I = A - P$

The formula method is good and short provided we can remember the formula.

 Example

Mr. Mofor invested the sum of 3,000,000 Frs. for 3 years at 10% per annum compound interest. Calculate his interest at the end of the 3 years using the compound interest formula.

Solution

$A = P\left(1 + \frac{r}{100}\right)^t = 3,000,000\left(1 + \frac{10}{100}\right)^3$
$= 3000000(1.1)^3 = 3,993,000$ Frs.

Compound interest $I = A - P$
$= 3,993,000 - 3,000,000 = 993,000$ Frs.

4.11 Compound Interest with Varying Principal

Sometimes, the principal increases because of increase in the loan or decreases due to partial repayment of the loan. In this case, we use the progressive method making sure that we add the extra loan to the total amount or subtract the partial repayment from the total amount at each instance before calculating the interest.

 Example

Mr. Nformi borrows 6,000,000 FCFA from his Credit Union at 3% per annum compound interest. He repays 2,000,000 FCFA at the end of each year. How much does he still owe at the end of the third repayment?

Module 5, Topic 4: Arithmetic Processes

Solution

Amount borrowed	= 6,000,000 FCFA
$I_1 = \frac{3}{100} \times 6,000,000$	=180,000 FCFA
Amount owed at end of year 1	= 6,180,000 FCFA
First repayment	= 2,000,000 FCFA
Amount owed after first repayment	= 4,180,000 FCFA
$I_2 = \frac{3}{100} \times 4,180,000$	= 125,400 FCFA
Amount owed at end of year 2	= 4,305,400 FCFA
Second repayment	= 2,000,000 FCFA
Amount owed after second repayment	= 2,305,400 FCFA
$I_3 = \frac{3}{100} \times 2,305,400$	= 69,162 FCFA
Amount owed at end of year 3	= 2,374,562 FCFA
Third repayment	= 2,000,000 FCFA
Amount owed after third repayment	= 374,562 FCFA

Therefore, the amount owed at the end of the third year is 374,562 FCFA.

 ## Exercise 4:5

1. Mr. Ngala borrowed 2,510,000 FCFA from his credit union at 10 % per annum compound interest. How much will he pay back at the end of 3 years?
2. Calculate the compound interest on 500,000 FCFA for 2 years at 6 % per annum.
3. Mrs. Teboh took a building loan of 7,000,000 FCFA. She pays a compound interest of 8% per annum. Given that she repairs 960,000 FCFA at the end of each year, how much does she still owe at the end of 3 years?
4. Miss Bih invests 100,000 FCFA on first January of each year at 5% per annum compound interest. Find to the nearest franc the total amount of her investment at the end of 3 years.
5. Mr. Ndumbe borrows 7,000,000 FCFA at 4% per annum compound interest. He repays 2,000,000 FCFA each year. In how many years will he clear loan?
6. A man invests 500,000 FCFA each year at 4% per annum compound interest. Calculate his investment just before he invests the fourth time.
7. A woman borrows 6,000,000 FCFA from a credit union at 3% per annum. She repays 2,000,000 FCFA at the end of each year. Calculate her debt after the

third repayment.

4.12 Currency Exchange- Decimal Currency

The currency of many countries is subdivided into units of 100 called cents. Such a currency is known as the decimal currency.

Some decimal currencies are shown in the following table.

Country	Basic unit	Sub unit
Europe	Euro (€)	Centime
Nigeria	Naira (₦)	Kobo (k)
USA	Dollars ($)	Cents (c)
Cameroon	Francs (FCFA)	-

In each case 100 subunits = 1 basic unit

Amounts less than the basic unit are expressed either as decimals to 2 decimal places or using the sub units. Thus,

$$28 \text{ p} = £ 0.28$$

$$200 \text{ naira, } 40 \text{ kobo} = ₦ 200.4$$

$$52 \text{ dollars, } 3 \text{ cents} = \$52.03$$

Conversion of Currency

International travelers must be able to interconvert currency because exchange rates though they exist, keep on fluctuating from time to time.

 Example

The rate of exchange is such that 1 FF = 120 FCFA and 45 Belgian Franc (BF) = 1 FF. Find in FCFA the cost of a motorcycle, which cost 180,000 BF.

Solution

45 BF = 1 FF

$\therefore 180,000 \text{ BF} = \dfrac{1}{45} \times 180,000 \text{ FF} = 4,000 \text{ FF}$

1 FF = 120 FCFA

$\therefore 4,000 \text{ FF} = 4,000 \times 120 \text{ FCFA} = 480,000 \text{ FCFA}$

Exercise 4:6

1. In September 1986, the exchange rates were as follows: $1 = 340 FCFA, £1 = $1.146. A student was required to pay £2000 as tuition for the 1986/87 academic year in a British university. Determine how much money was required in FCFA.
2. The rate of exchange is such that 1FB = 7.83 FCFA and $1 = 313.52 FB. Find in FCFA, to the nearest thousand francs, the cost of a car marked at $1000.
3. An Englishman bought a car from France for 90,000 FF. The exchange rate was then, £1= 11.77 FF. He paid for the car when the exchange rate was £1= 16.68 FF. How much did he lose to the nearest £?

Multiple Choice Exercise 4

1. 8 T-shirts are bought for 20,000 FCFA. The cost of 7 T-shirts in FCFA will be:
   [A] 13000   [B] 15000   [C] 17500   [D] 18500
2. 9 boys can fill a tank in 4 hours. How many such boys will be required to fill a similar tank in 9 hours:
   [A] 3   [B] 6   [C] 4   [D] 9
3. Given that 20 men take 6 days to clear a field. The time it would take 12 men working at the same rate to clear a similar field is:
   [A] 40 days   [B] 2 days   [C] $3\frac{1}{2}$ days   [D] 10 days
4. A car travels 219 km using 27 litres of gas. At that rate, how far can the car travel using 162 litres of gas?
   [A] 1314 km   [B] 816 km   [C] 952 km   [D] 1088 km
5. A van travels 180 km on 6 litres of gas. How many litres will it need to travel 750 km?
   [A] 75 litres   [B] 25 litres   [C] 50 litres   [D] 225 litres
6. The cost of 4 articles at 720 FCFA each is:
   [A] 2920 FCFA   [B] 2880 FCFA   [C] 2820 FCFA   [D] 2900 FCFA
7. The cost of 2.5 kg of tomatoes at 360 FCFA per kg is:
   [A] 900 FCFA   [B] 860 FCFA   [C] 1100 FCFA   [D] 720 FCFA
8. The cost of 2 metres of material at 1200 FCFA per metre and 3 metres at 1500 FCFA per metre is:
   [A] 6900 FCFA   [B] 13500 FCFA   [C] 6600 FCFA   [D] 7100 FCFA
9. The cost of 2000 articles at 25 FCFA each is:
   [A] 5,000 FCFA   [B] 25,000 FCFA   [C] 50,000 FCFA   [D] 500,000 FCFA
10. The change from 1000 FCFA after buying 18 buttons at 5 FCFA each is:
    [A] 990 FCFA   [B] 890 FCFA   [C] 810 FCFA   [D] 910 FCFA

11. The cost of 200 articles at 30 FCFA each is:
    [A] 7000 FCFA [B] 6000 FCFA [C] 5000 FCFA [D] 5700 FCFA
12. A furniture maker estimated that the cost of making a cupboard would be about 250000 FRS. He bought the materials and found that the cost came to 275000 FRS. The percentage increase in the estimate is:
    [A] 5% [B] 10% [C] 15% [D] 20%
13. For his holidays, a man put aside 10% of his 15000 FCFA weekly wage for 40 weeks in the year. The amount saved for his holidays is:
    [A] 60000 FCFA [B] 30000 FCFA [C] 15000 FCFA [D] 150000 FCFA
14. A girl bought a record for 1500 FCFA and sold it for 1200 FCFA. Her loss as a percentage of the cost price is:
    [A] 15% [B] 20% [C] 60% [D] 75%
15. A woman bought a dress for 15000 FCFA. To make a profit of 20% the dress should be sold at:
    [A] 19,000 FRS [B] 18,000 FRS [C] 16,000 FRS [D] 14,400 FRS
16. A trader made a profit of 50% on his cost price by selling a radio at 15,000 FCFA. The record cost him:
    [A] 5,000 FRS [B] 7,500 FRS [C] 10,000 FRS [D] 25,000 FRS
17. This year the budget of a development association decreased by 4 % given that its annual budget for last year was 32,000,000 FRS. Its budget for this year is:
    [A] 30,720,000FRS [B] 25,000FRS [C] 33,280,000FRS [D] 2,240,000FRS
18. By selling some crates of soft drinks for 6000 FRS, a dealer makes a profit of 50 %. The amount which the dealer pays for the drinks is:
    [A] 12,000 FRS [B] 25,000 FRS [C] 4,500 FRS [D] 4,000 FRS
19. A trader makes a loss of 15 % when selling an article.
    The ratio selling price : cost price is:
    [A] 3:20 [B] 3:17 [C] 17:20 [D] 20:23
20. A man made a loss of 15 % by selling an article for 59500 FRS. The cost price of the article was:
    [A] 60,000 FRS [B] 70,000 FRS [C] 68,425 FRS [D] 89,250 FRS
21. Mr. Anyang bought a piece of land for 2.5 million FCFA and sold it for 3 million FCFA. His percentage profit is:
    [A] $18\frac{2}{3}\%$ [B] 20% [C] 16.7% [D] 25%
22. If the simple interest on 200,000 FRS after 9 months is 6000 FRS, the interest rate per annum is:
    [A] $2\frac{1}{4}\%$ [B] 6% [C] 5% [D] 4%
23. A cooperative society charges an interest of $5\frac{1}{2}\%$ per annum on any amount borrowed by its members. If a member borrows 125000 FRS. The amount he pays back after 1 year will be:
    [A] 136,875 FRS [B] 131,875 FRS [C] 128,750 FRS [D] 126,250 FRS
24. The compound interest on 400,000 FRS for 2 years at 8 % per annum is:
    [A] 32,000 FRS [B] 34,560 FRS [C] 66,560 FRS [D] 43,200 FRS
25. The exchange rate for FCFA is 1000 FCFA for 238 Naira. 2,856,000 FCFA changed into Naira would be:
    [A] 12 [B] 1200 [C] 679728 [D] 6797280
26. 33. Given that 1000 FCFA = ₦ 240. ₦ 30,000 will be equivalent to:

[A] 250,000 FCFA [B] 80,000 FCFA [C] 125,000 FCFA [D] 72,000 FCFA
27. Given that $3 = 720 FCFA. 600,000 FCFA in dollars ($) is:
    [A] $2,500   [B] $14,400,000   [C] $277   [D] $200 00
28. At the current exchange rate, 1 dollar ($) = 500 FCFA and 1 pound = 1000 FCFA. 200 pounds, exchanged to dollars will be:
    [A] $400   [B] $10.00   [C] $100   [D] $2.00

Topic 5

REAL NUMBERS

Objectives

At the end of this topic, the learner should be able to:

1. Define and represent the set of real numbers.
2. Carry out operations in $\mathbb{R}$.
3. Compare real numbers using < and >.
4. Represent numbers on a real number line.
5. Represent intervals on a real number line.

Module 5, Topic 5: Real Numbers

5.1 Radicals

The square root, cube root and other roots of numbers are called radicals.

You are advised to go to sections 3.9 and 3.11 of book 1 and revise the method of finding square roots and cube roots.

Skill Building Exercise 5:1

Evaluate the following
(a) $\sqrt{2500}$ (b) $\sqrt{400}$ (c) $\sqrt{576}$ (d) $\sqrt{1089}$ (e) $\sqrt{1225}$
(f) $\sqrt[3]{729}$ (g) $\sqrt[3]{1728}$ (h) $\sqrt[3]{5832}$ (i) $\sqrt[3]{2744}$ (j) $\sqrt[3]{3375}$

5.2 Irrational Numbers $\mathbb{Q}'$

In topic 3, we saw that numbers can be expressed as either whole numbers or terminating decimals or recurring decimals.

Brainstorming Exercise

1. Use a calculator to evaluate the following.
    (a) $\sqrt{36}$ (b) $\sqrt{50}$ (c) $\sqrt{-36}$
2. Do all the numbers have exact square roots?
3. What result does the calculator give you for $\sqrt{-36}$?
4. Is it all numbers which have this property that they can be expressed as either whole numbers or terminating decimals or recurring decimals?
5. Which are those numbers that cannot be expressed in this way?

Many numbers are not whole numbers and cannot be expressed as terminating or recurring (repeating) decimals. One of these sets of numbers is the set of **irrational numbers** denoted by $\mathbb{Q}'$. Irrational numbers are numbers that cannot be expressed as the quotient or ratio of two integers. The square root of a whole number is an irrational number if the whole number cannot be expressed as a perfect square.

Examples of irrational numbers are $\sqrt{2}, \sqrt{3}, \sqrt{7}, \pi$ etc.

Note! The symbol $\sqrt{}$ means 'the positive square root of'. Thus $\sqrt{4} = 2$, $\sqrt{9} = 3$, $\sqrt{25} = 5$ etc.

It should be noted that whereas natural numbers can be represented as exact points on a number line, irrational numbers can only be represented as approximate points.

Competency Base Mathematics for Secondary Schools Book 2

Exercise 5:1

Which of the following are irrational numbers?
1. $\sqrt{2}$
2. $\sqrt{49}$
3. $\sqrt{50}$
4. $\sqrt{21}$
5. $\sqrt{81}$
6. $\sqrt{71}$
7. $\sqrt{64}$
8. $\sqrt{16}$
9. $\sqrt{60}$
10. $\sqrt{84}$

5.3 Real Numbers, $\mathbb{R}$

All the rational numbers and all the irrational numbers put together constitute the set of real numbers denoted by $\mathbb{R}$. Numbers such as $\sqrt{-36}$, are not real numbers. It can thus be said that, **all natural numbers are integers, all integers are rational numbers** and **all rational numbers** and **all irrational numbers are real numbers**.

The statement "all natural numbers are integers" is written symbolically as $\mathbb{N} \subset \mathbb{Z}$. WE can also write this statement as $\mathbb{Z} \supset \mathbb{N}$. Using this notation we can write the mathematical statement $\mathbb{N} \subset \mathbb{Z} \subset \mathbb{Q}$ to show the relationship between the three sets of numbers.

Use $\supset$ to write a statement which means the same as the above. Other ways of reading the statement $\mathbb{N} \subset \mathbb{Z}$ are "$\mathbb{N}$ is contained in $\mathbb{Z}$" or "$\mathbb{N}$ is a subset of $\mathbb{Z}$".

We can write the statement '3 is a natural number' symbolically as $3 \in \mathbb{N}$. Write a mathematical statement which reads "$\frac{1}{2}$ is a rational number".

5.4 The Real Number Line

Any real number can be represented by a point on a real number line and every point on a real number line represents a real number. On a number line, the point which is to the left represents a smaller number and the point which is to the right represents a larger number.

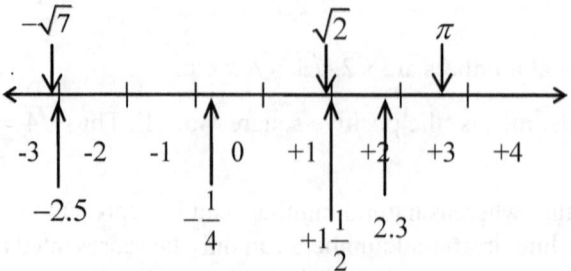

Exercise 5:2

1. Mark the points $-\frac{10}{3}$, $\frac{13}{4}$, 2, 0, -2.5, $\sqrt{13}$

2. On the number line below,
   (a) π should be plotted as a point between _____
   (b) $-\frac{19}{5}$ should be plotted as a point between _____
   (c) $0.2\dot{4}$ should be plotted as a point between _____
   (d) $\frac{13}{3}$ should be plotted as a point between _____
   (e) $\sqrt{7}$ should be plotted as a point between _____
   (f) $-\sqrt{18}$ should be plotted as a point between _____

3. Five numbers A, B, C, D and E are such that C is greater than A but less than E. E is less than B. D is greater than B. Arrange A, B, C, D and E in order as they would be plotted on a number line.

5.5 Relationship between Sets of Numbers

The relationship between the various sets of real numbers discussed above can be illustrated by a Venn diagram or tree diagram as follows.

(a)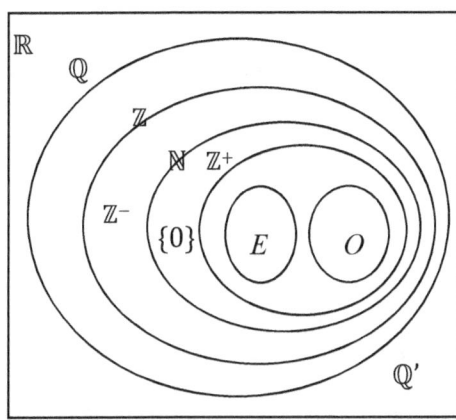

(b)

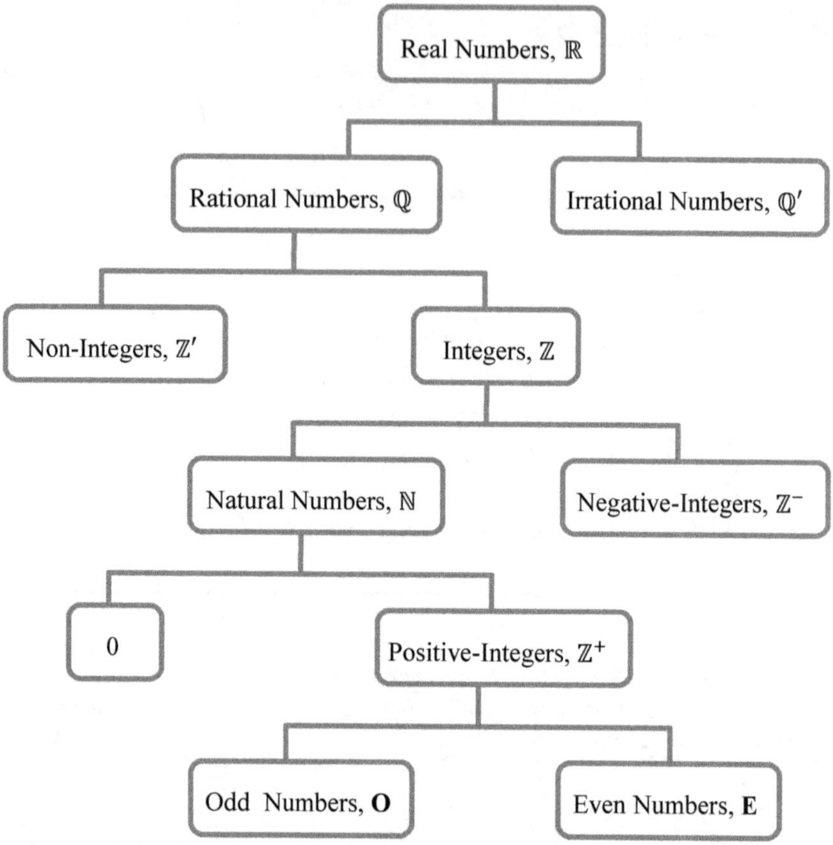

```
                        Real Numbers, ℝ
                       /              \
          Rational Numbers, ℚ      Irrational Numbers, ℚ'
           /            \
   Non-Integers, ℤ'    Integers, ℤ
                       /        \
              Natural Numbers, ℕ    Negative-Integers, ℤ⁻
               /        \
              0      Positive-Integers, ℤ⁺
                       /        \
              Odd Numbers, O    Even Numbers, E
```

From the Venn diagram in figure (a) above or the tree diagram in figure (b), we can see that all natural numbers are integers and all integers are rational numbers.

 Exercise 5:3

1. Which of the following numbers are rational and which are irrational?
   (a) 9 (b) $\sqrt{25}$ (c) $\sqrt{7}$ (d) π (e) $\sqrt{9}$ (f) $\sqrt{11}$
2. Represent the following rational numbers on a number line.
   $+5, -3, 4.5, 3\frac{1}{4}, -2\frac{3}{4}, 2.7$.
3. List the integers between -10 and $+10$ inclusive.
4. How many rational numbers are there between -2 and $+2$? Give an explanation for your answer.
5. State two rational numbers that lie between the following pairs of rational numbers.
   (a) -3.051 and -3.052 (b) $+5.159$ and $+5.2$

6. Use the symbols <, > or = to compare the following rational numbers.

   (a) $\dfrac{2}{5} \ldots \dfrac{3}{8}$ (b) $-\dfrac{1}{2} \ldots -\dfrac{1}{3}$ (c) $\dfrac{1}{2} \ldots \dfrac{1}{3}$

   (d) $-3.54 \ldots 3.549$ (e) $0.57 \ldots \dfrac{4}{7}$ (f) $-\dfrac{2}{3} \ldots -\dfrac{4}{5}$

   (g) $\dfrac{2}{3} \ldots \dfrac{4}{5}$ (h) $-\dfrac{7}{9} \ldots -0.779$ (i) $\dfrac{11}{13} \ldots \dfrac{23}{25}$

7. Represent each pair of rational numbers in question 6 on a real line.
8. Classify the numbers $\dfrac{2}{3}, -5.3, \pi, -\sqrt{3}, 1\dfrac{1}{2}, -\dfrac{4}{11}$ under the sets in the table below. One number may belong to more than one set.

$\mathbb{N}$	$\mathbb{Z}$	$\mathbb{Q}$	$\mathbb{Q}'$	$\mathbb{R}$

9. Fill in the blanks using $\mathbb{N}$, $\mathbb{Z}$, $\mathbb{Q}$ or $\mathbb{Q}'$.
   (a) All the numbers that belong to _____ also belong to $\mathbb{Z}$.
   (b) All the numbers that belong to $\mathbb{Z}$ also belong to _____.
   (c) The sets _____ and _____ have no elements in common.
   (d) The set $\mathbb{R}$ of real numbers is made up of all the elements of _____ and _____ put together.

5.6 Intervals

Recall that a statement that two real quantities or expressions are not equal is called an **inequality**.

    < means 'is less than'

    > means 'is greater than'

    ≤ means 'is less than or equal to'

    ≥ means 'is greater than or equal to'

5.7 Ordering

$\forall a, b, c \in \mathbb{R}$,

1. Either $a < b$ or $a > b$ or $a = b$.
2. $a < b \Leftrightarrow b > a$.
3. If $a < b$ and $b < c$ then $a < c$. This is known as the **transitive property** of inequalities.
4. If $a < b$ and $c > 0$ then $ac < bc$ and if $a < b$ and $c < 0$ then $ac > bc$.

5.8 Representation of Intervals

On a real number line, if $a < b$ then the point a is to the left of the point b. When representing inequalities on a number line an open circle o is used when the boundary point is not included and a fill-in circle ● is used when the boundary point is included.

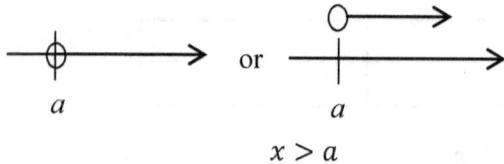

$x > a$

The boundary point a is not included.

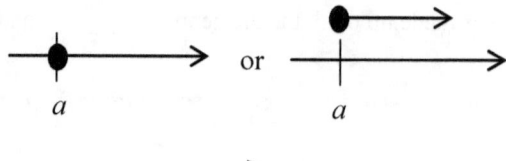

$x \geq a$

The boundary point a is included.

Open Intervals

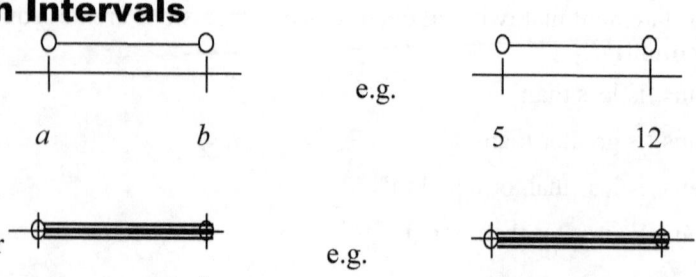

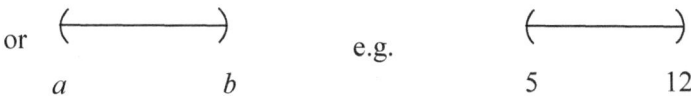

The boundary points a and b are not included.

Closed Intervals

In closed intervals the boundary points a and b are included. This is denoted by $a \leq x \leq b$ or $[a, b]$. For instance $-4 \leq x \leq 5 = [-4, 5]$. This is represented as

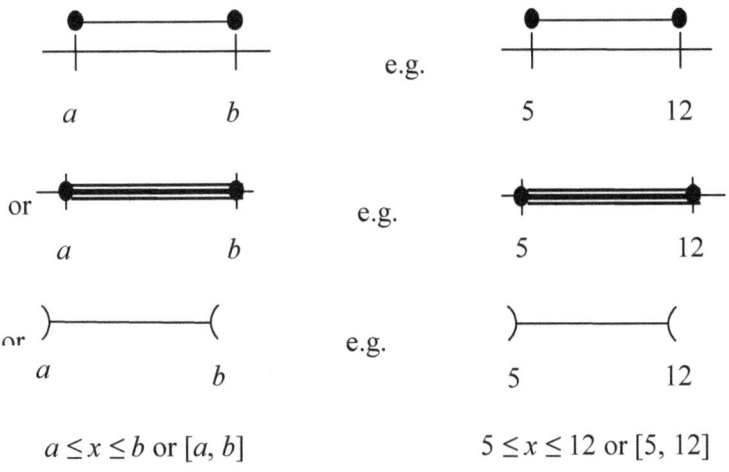

$a \leq x \leq b$ or $[a, b]$ $5 \leq x \leq 12$ or $[5, 12]$

The boundary points a and b are included.

Half Open or Half Closed Intervals

Closed on the right or Open-Closed Interval

In an interval open to the left and closed to the right the left boundary point a is excluded and the right boundary point b is included. This is denoted by $a < x \leq b$ or $(a, b]$ or $]a, b]$.

Thus, $5 < x \leq 12 =]5, 12] = (5, 12]$. This is represented as

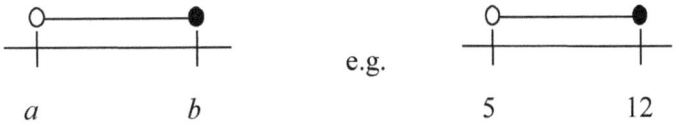

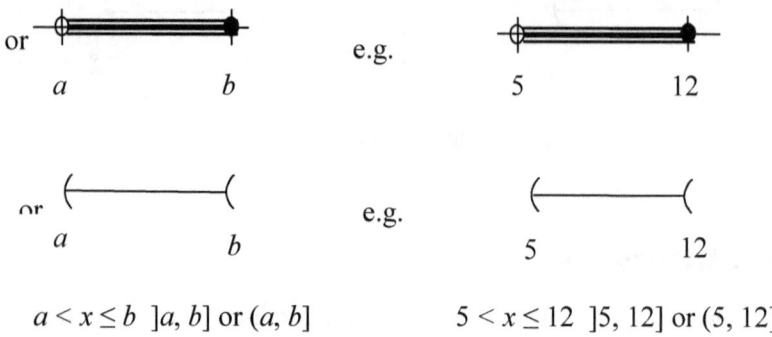

$a < x \leq b$]a, b] or (a, b] $5 < x \leq 12$]5, 12] or (5, 12]

The boundary point a is not included but b is included.

Closed on the left or Closed- Open Interval

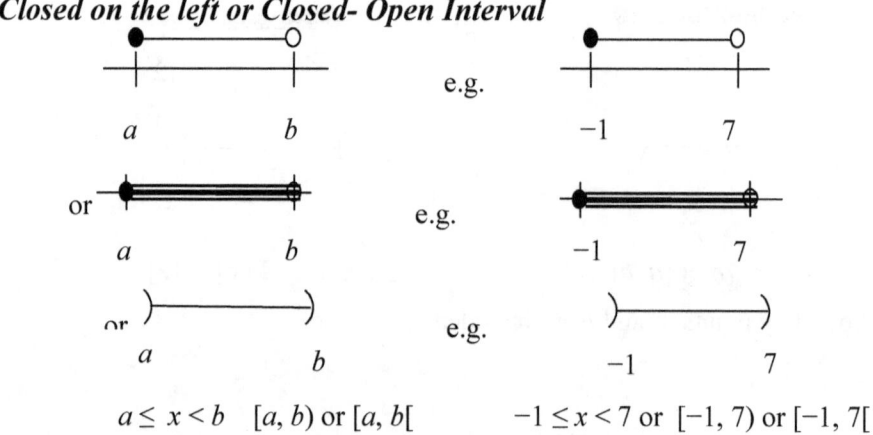

$a \leq x < b$ [a, b) or [a, b[$-1 \leq x < 7$ or [−1, 7) or [−1, 7[

The boundary point a is included but b is not included.

Notice that three different notations associated with closed and opened intervals have been used. The reader is required to be able to recognize and use these notations but in each instance should use only one of them.

Module 5, Topic 5: Real Numbers

Exercise 5:4

1. Represent each of the following on a number line.
   (a) (2,0)   (b)]4,1[(c) [3,1]
   (d) $\{x : 3 < x < 8, x \in \mathbb{R}\}$ (e) $\{x : 4 \leq x \leq 8, x \in \mathbb{R}\}$

2. Name the intervals represented on the following number lines.

   (a)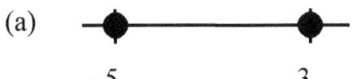

   (b)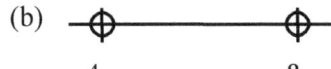

   (c)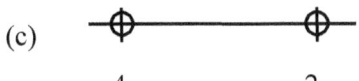

   (d)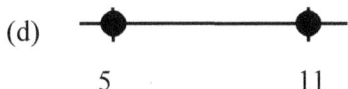

   (e)

   (f)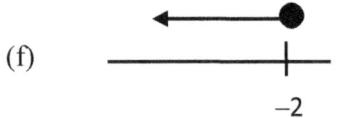

3. Classify the following intervals as closed, open or half open, half closed intervals.

   (a)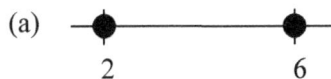

   (b)

   (c) (4,6)   (d) [7,3]   (e) [2,4[
   (f) $-3 < x < 8$ (g) $\{x : -2 \leq x \leq 0, x \in \mathbb{R}\}$

4. Represent each of the following intervals on a real number line.
   (a) (−2,5]   (b) (−10,−6)   (c) [0,8]
   (d)]−3,9[(e) $\{x : -11 < x \leq 5, x \in \mathbb{R}\}$

5. State which of the following intervals are closed, open or half open (half closed) giving reasons for your answer.
   $P = \{17 \leq x \leq 17, x \in \mathbb{R}\}$, $Q = [1,3]$, $R = (-1,2]$,
   $S = \{-2 \leq x \leq 2, x \in \mathbb{R}\}$, $T = [2,5]$, $U =]-2,0[$,
   $V = [5,8]$, $W = \{-7 < x < -2, x \in \mathbb{R}\}$.

6. Represent each of the intervals in question 5, above on a number line.

7. Give the other notations for each of the following intervals.
   (a) $\{x : -2 < x \leq 2\}$ (b) [98, 101]   (c) (2,5)   (d) $\{x : 1 \leq x < 3\}$
   (e)]−6, −1[(f) (14,19]   (g) [0,7)   (h) [−3,1[(i)]11,13]

Competency Base Mathematics for Secondary Schools Book 2

 Multiple Choice Exercise 5

1. The value of 136×47 is 6392. The value of $\frac{63.92}{13.6}$ is:
   [A] 47   [B] 0.047   [C] 0.47   [D] 4.7
2. Given that $225 \times 35 = 7875$, then $22.5 \times 0.35 = 7875$ is equal to:
   [A] 0.07875   [B] 0.7875   [C] 7.875   [D] 78.75
3. The value of $\frac{1}{0.2} + \frac{1}{0.25}$ is:
   [A] 45   [B] 4.5   [C] 2.5   [D] 9
4. $78.75 \div 0.35$ is:
   [A] 0.225   [B] 0.25   [C] 22.5   [D] 225
5. The reciprocal of 0.02 is:
   [A] 500   [B] 50   [C] 0.5   [D] 0.05
6. The reciprocal of 0.0002 is:
   [A] 50   [B] 500   [C] 5000   [D] 50,000
7. To one decimal place, the reciprocal of 0.625 is:
   [A] 1.6   [B] 0.6   [C] 6.3   [D] 62.5
8. $5 - 0.003$ equals:
   [A] 0.002   [B] 4.003   [C] 4.007   [D] 4.997
9. The value of $4.7 - 1.9 + 2.1$ is:
   [A] 5.9   [B] 8.7   [C] 1.7   [D] 4.9
10. The value of 0.6×0.04 is:
    [A] 0.24   [B] 0.64   [C] 0.024   [D] 2.4
11. 0.4×1.4 equals:
    [A] 56   [B] 5.6   [C] 0.56   [D] 0.056
12. 0.4×0.2 equals:
    [A] 0.8   [B] 0.08   [C] 8   [D] 0.6
13. $0.93 + 0.08$ is equal to:
    [A] 10.1   [B] 101   [C] 1.01   [D] 0.101
14. $0.1 \times 0.2 \times 0.3$ is equal to:
    [A] 0.06   [B] 0.006   [C] 0.05   [D] 0.005
15. The value of $\frac{2.4}{4}$ is:
    [A] 0.6   [B] 6   [C] 60   [D] 2.1
16. The result of squaring the number 6 is:
    [A] 12   [B] 26   [C] 36   [D] 62
17. $\sqrt{7744}$ in index form is:
    [A] 26   [B] $2^3 \times 13$ [C] $2^3 \times 9$ [D] $2^3 \times 11$
18. $\sqrt{3136}$ in index form is:
    [A] 2×7^2 [B] $2^3 \times 7$ [C] $2^2 \times 7$ [D] 2×7
19. The smallest number to multiply by $3^2 \times 5$ to give a perfect square is:
    [A] 5   [B] 6   [C] 15   [D] 25
20. The least number which multiplies 54 to make a perfect square is:

Module 5, Topic 5: Real Numbers

[A] 3   [B] 4   [C] 6   [D] 8

21. It is true to say that:
 [A] $\sqrt{5} \in \mathbb{R}$ [B] $\sqrt{2} \in \mathbb{Q}$ [C] $-5 \in \mathbb{N}$ [D] $3 \in \mathbb{Q}'$

22. It is not true to say that:
 [A] $16 \in \mathbb{N}$ [B] $\sqrt{9} \in \mathbb{R}$ [C] $\pi \in \mathbb{R}$ [D] $\frac{1}{3} \in \mathbb{Z}$

23. It is not true to say that:
 [A] $-2 \in \mathbb{Z}$ [B] $7 \in \mathbb{Z}$ [C] $\pi \in \mathbb{Q}$ [D] $3 \in \mathbb{Q}$

24. It is true to say that:
 [A] $-5 \in \mathbb{Q}$ [B] $-5 \in \mathbb{N}$ [C] $\frac{1}{3} \in \mathbb{N}$ [D] $\frac{1}{3} \in \mathbb{Z}$

25. It is not true to say that:
 [A] $-5 \in \mathbb{Q}$ [B] $16 \in \mathbb{N}$ [C] $-16 \in \mathbb{Z}$ [D] $\frac{1}{3} \in \mathbb{Z}$

26. It is not true to say that:
 [A] $\mathbb{N} \subset \mathbb{Z}$ [B] $\mathbb{N} \subset \mathbb{Q}$ [C] $\mathbb{Q} \subset \mathbb{N}$ [D] $\mathbb{Z} \subset \mathbb{Q}$

27. The number which does not belong to the set $\mathbb{Z}$ of integers is:
 [A] -14 [B] $\frac{1}{2}$ [C] 32   [D] 0

28. The number which belongs to all the three sets $\mathbb{Z}$, $\mathbb{Q}$, and $\mathbb{N}$ is:
 [A] 9   [B] -7 [C] $\frac{7}{9}$ [D] $-\frac{7}{9}$

29. The correct statement is:
 [A] $\mathbb{Z} \subset \mathbb{Q} \subset \mathbb{N}$ [B] $\mathbb{Q} \subset \mathbb{N} \subset \mathbb{Z}$ [C] $\mathbb{N} \subset \mathbb{Z} \subset \mathbb{Q}$ [D] $\mathbb{N} \subset \mathbb{Q} \subset \mathbb{Z}$

30. The correct Venn diagram below is:

 [A] 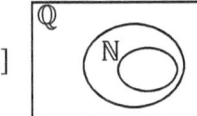   [B]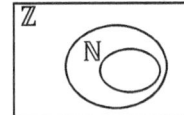

 [C] 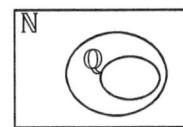   [D]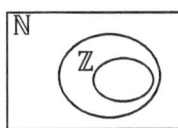

31. If x varies over the set of real numbers, the inequality illustrated in the number line below is:
 [A] $-3 < x \leq 2$ [B] $-3 \leq x < 2$ [C] $-3 < x < 2$ [D] $-3 \leq x \leq 2$

32. If x is a real number the inequality represented in the figure below is:
 [A] $\{x: -5 < x \leq 3\}$ [B] $\{x: -5 \leq x \leq 3\}$
 [C] $\{x: -5 \leq x < 3\}$ [D] $\{x: -5 < x < 3\}$

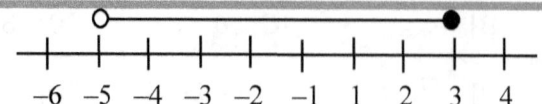

33. If $x \in \mathbb{R}$, the inequality more illustrated in the number line below is:
    [A] $x < 4$ [B] $x > -2$ [C] $-2 < x \leq 4$ [D] $-2 \leq x < 4$

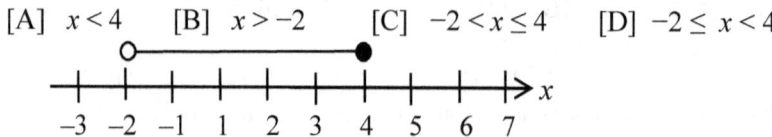

34. If x is real, the inequality more illustrated in the number line below is:
    [A] $x < 4$ [B] $x > -2$ [C] $-2 < x \leq 4$ [D] $-2 \leq x < 4$

35. If x varies over the set of real numbers, the inequality illustrated below is:
    [A] $-2 \leq x < 3$ [B] $-2 < x \leq 3$ [C] $-2 < x < 3$ [D] $-2 \leq x \leq 3$

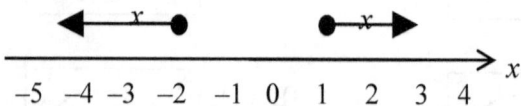

36. The pairs of inequalities represented on the number line below are:
    [A] $x < -2$ and $x \geq 1$ [B] $x \leq -2$ and $x > 1$
    [C] $x \leq -2$ and $x < 1$ [D] $x \leq -2$ and $x \geq 1$

37. The number line below which represents the inequality $2 \leq x < 9$ is:

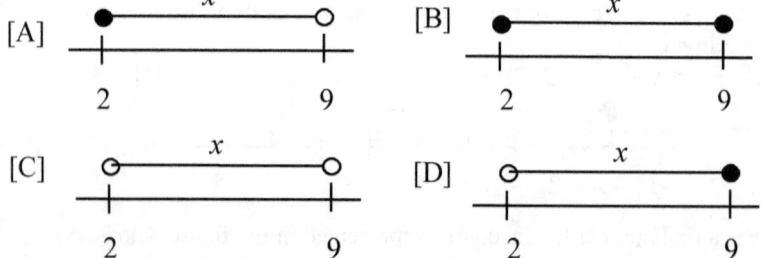

38. The number line below represents:
    [A] $0 < x < -7$ [B] $-7 < x < 1$ [C] $-7 < x \leq -1$ [D] $-7 \leq x < -1$

Module 6

Introduction to Plane Geometry

Family of Situations

Module 6 is an extension of module 2 and at the end of the module; the student is expected to acquire many more competencies within the **families of situations** *'Representation and transformation of Plane Shapes within the Environment'*.

Categories of Action

The categories of action for module 6 include:
1. Recognition of plane shapes within the physical environment,
2. production of plane shapes and transformation of the physical environment
3. Determination of measures and position within the physical environment.

Credit

The module is expected to be covered within 10 weeks teaching 4 hours per week (or within 40 hours).

Topic 6

DISTANCES

Objectives

At the end of this topic, the learner should be able to:

1. Determine the horizontal distance between two points on a number line.
2. Determine the vertical distance between two points on a number line.
3. Construct the mediator of a line segment.

Module 6, Topic 6: Distances

Review Exercise

1. How do we denote the length of a line segment between the points A and B?
2. Explain how you would measure the length of a line segment using a pair of dividers.
3. Measure the length of the line segments indicated on the following number line using a pair of dividers and a ruler. (a) AB (b) BC (c) AC

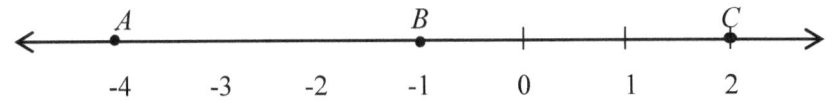

Recall that the length of a line segment between the points A and B is denoted by $d(AB)$ or simply AB. The length or space between any two points such as AB and AC in the above figure is called the distance between the two points and is denoted in the same way.

6.1 Horizontal or Vertical Distance between two Points

On the number line above, the distance between the points A and C can be obtained by counting the number of units between A and C. Thus, $AC = 6$ units

Example

Using the number line below, find

(a) AB (b) BD (c) AC (d) AD (e) AB + CD

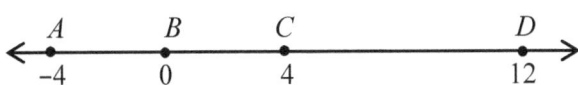

Solution
(a) $AB = 4$ units (b) $BD = 12$ units (c) $AC = 8$ units
(d) $AD = 16$ units (e) $AB + CD = 12$ units.

Exercise 6:1

1. Using the number line below, find
   (a) AB (b) BD (c) AC (d) AE (e) AC + ED

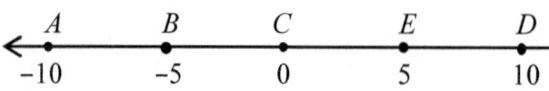

4. Measure the length of the following line segments.

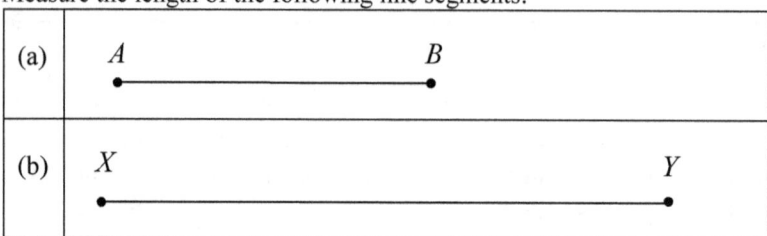

2. Using the number line in figure (i), find
   (a) RP (b) SQ (c) RQ (d) SP (e) SR+QP

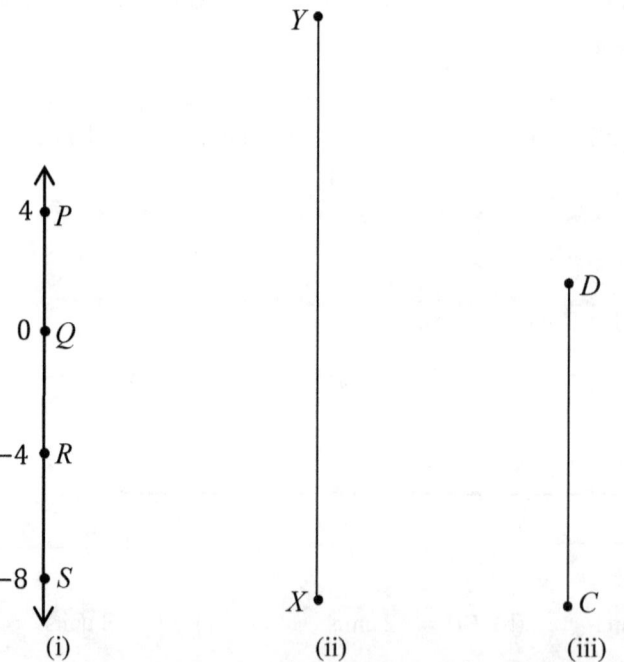

(i)　　　　　　(ii)　　　　　　(iii)

5. Measure the length of the line segments in figure (ii) and (iii) above.
6. Two points P and Q on a line have coordinates −17.5 and +8 respectively. Find the distance PQ.
7. A fly moves on a line from the point with coordinate is 13.3 to a point with coordinate −10.6. Find the distance covered by the fly.

Module 6, Topic 6: Distances

8. On the real number line in below, determine the distance between:
   (a) PQ   (b) PR   (c) RS   (d) PS   (e) QS

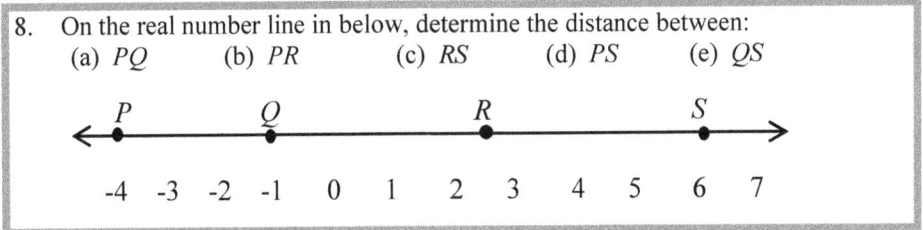

6.2 Midpoint and Mediator of a Line Segment

The midpoint of a line segment is the point that divides the line segment into two equal segments. In the figure below, *M* is the midpoint of the line segment *AB*. If a perpendicular is drawn to the line *AB* to pass through its midpoint *M*, this perpendicular is called the mediator or perpendicular bisector of *AB*.

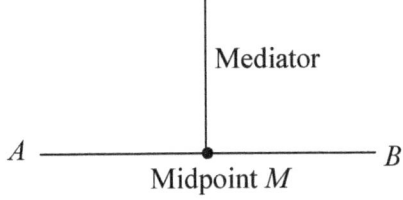

Generally, if **M** ∈ [**AB**] such that **AM** = **MB** and **AM** + **MB** = **AB**, then *M* is the midpoint of [**AB**].

Conversely, if *M* is the midpoint of [*AB*], then *AM* + *MB* = *AB*.

The **mediator** or **perpendicular bisector** of a given line segment is a line which is perpendicular and passes through the midpoint of the line as shown in the figure above.

6.3 Constructing the Mediator of a Line Segment

Example

Construct the perpendicular bisector of the line segment *AB*.

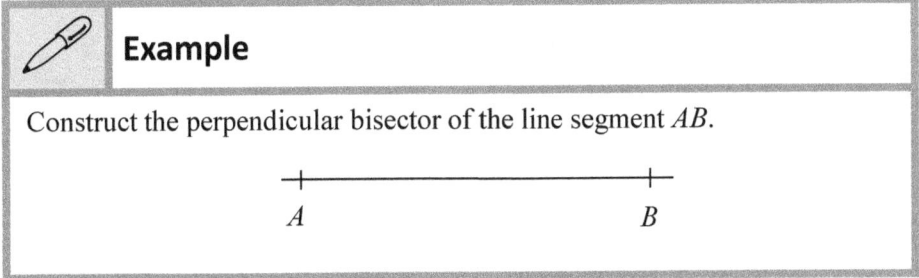

Procedure

(i) With centre A, draw two arcs of equal radii on the two sides of AB.
(ii) With centre B draw two arcs of equal radii as those in (a) to cut the first two at C and D.
(iii) Now join the points C and D of intersection of these arcs with ruler and pencil. The line CD is the perpendicular bisector of the line segment AB.

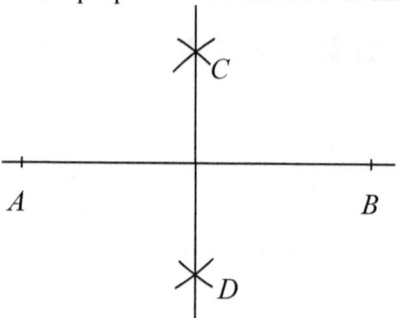

 Exercise 6:2

1. In the figure below, $[XY]$ is a line segment. Mark a point M where M is the midpoint of $[XY]$.

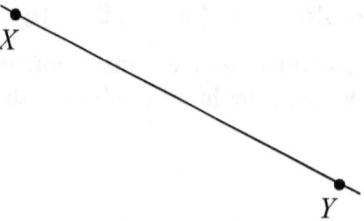

2. Draw a straight-line segment $[CD]$ of length 8 cm. Bisect the line $[CD]$.

Multiple Choice Exercise 6

1. *AB* bisects *PQ* at point *N*. The statement that is true of *N* is:
   [A] *N* is the midpoint of AB.
   [B] *N* is the midpoint of *AB* and the midpoint of *PQ*.
   [C] *N* is the midpoint of *PQ*.
   [D] *N* divides *PQ* in the ratio 2:1.
2. Point *P* is the midpoint of *AB*. Complete the statement: *PB* = 7 cm, *AB* is equal to:
   [A] 7 cm   [B] 14 cm   [C] 3.5 cm   [D] none of the above
3. The name of the instrument in the figure below is:
   [A] a protractor   [B] a pair of dividers
   [C] a pair of compass   [D] a set square

4. In order to construct a perpendicular bisector one needs at least:
   [A] a pair of compass, pencil and ruler
   [B] a protractor, pencil and ruler
   [C] a set square, pencil and ruler
   [D] a protractor, pencil and a set square
5. The name of the instrument in the figure below is:
   [A] a set square   [B] a pair of compasses
   [C] a pair of dividers   [D] a protractor

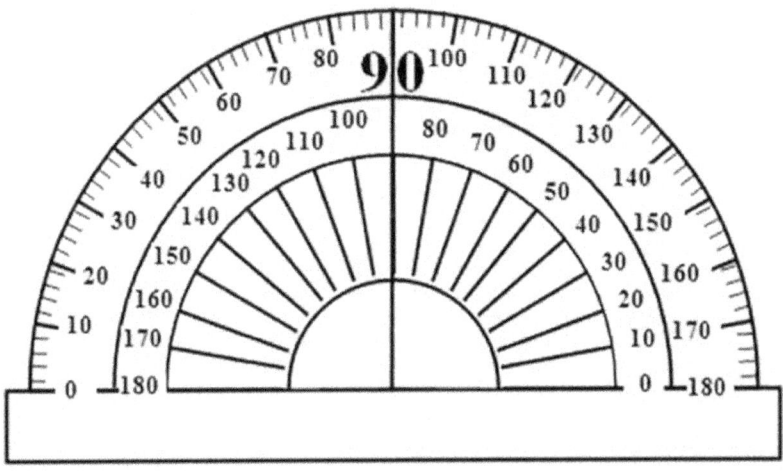

6. The path of a moving point such that its distance from a fixed line *l* is always constant, is:
   [A] a mediator of *l*　　　　　　　[B] a line parallel to *l*
   [C] a parabola　　　　　　　　　　[D] a circle
7. The path of a moving point such that its distance from two fixed points is always equal, is:
   [A] a perpendicular bisector *l*.　　[B] a line parallel to *l*.
   [C] the angle bisector from the two points.　　[D] a circle.
8. The diagram below which shows the construction of a perpendicular bisector is:

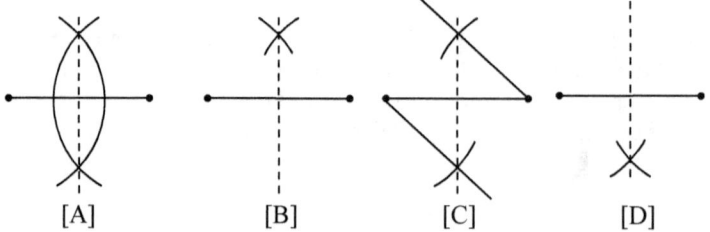

　　[A]　　　　　[B]　　　　　[C]　　　　　[D]

9. Another name for perpendicular bisector is:
   [A] Altitude　　　　　　　　　　　[B] Midpoint
   [C] Angle bisector　　　　　　　　[D] Mediator
10. In the figure below, *AM* is said to be:
    [A] The median　　　　　　　　　[B] The mediator
    [C] The altitude　　　　　　　　　[D] The angle bisector

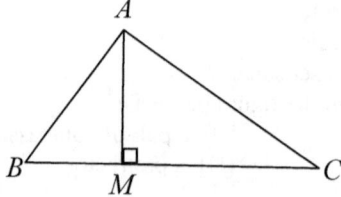

11. Given that in the following figure, M is the midpoint of *BC*, *AM* is called:
    [A] The median　　　　　　　　　[B] The mediator
    [C] The altitude　　　　　　　　　[D] The angle bisector

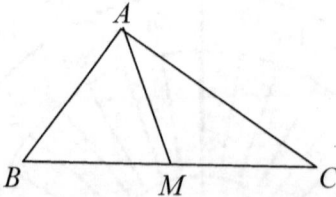

12. Given that in the following figure, *M* is the midpoint of *BC*, we call *NM*:
    [A] The median　　　　　　　　　[B] The mediator
    [C] The altitude　　　　　　　　　[D] The angle bisector

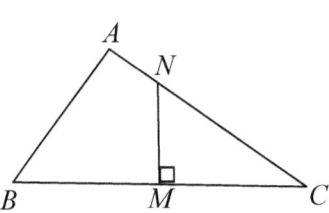

13. The following figure shows the construction of:
    [A] A mediator [B] A perpendicular to AB
    [C] An angle bisector [D] Intersecting lines

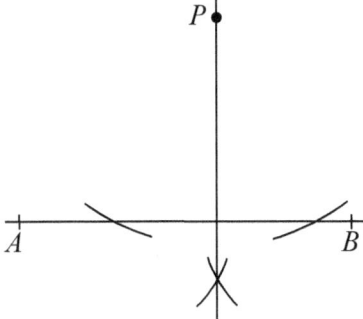

Topic 7

ANGLES

Objectives

At the end of this topic, the learner should be able to:

1. Determine angles between intersecting lines.
2. Determine vertically opposite angles.
3. Angles formed by two lines and a transversal.
4. Angles formed by parallel lines and a transversal.
5. Determine corresponding angles, alternate angles, co-interior angles.
6. Determine the properties of alternate angles.

Module 6, Topic 7: Angles

7.1 Angles

In Topic 10 of book 1, an angle was defined as the measure of the amount of turn of an object. Different types of angles such as; a full turn or a revolution, a half turn, a quarter turn or a right angle, an acute angle, an obtuse angle, a reflex angle were discussed. The standard notation for angles and some angles formed by intersecting straight lines such as, adjacent angles, vertically opposite angles and angles on a straight line were also examined. The learner is advised to review this material before proceeding to the next section of this Topic.

7.2 Parallel Lines and Transversals

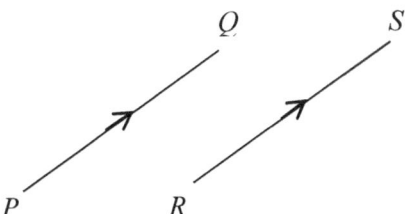

Two lines are said to be **parallel** if they have the same direction. Parallel lines never meet. The figure above shows, that the lines PQ and RS are parallel.

A **transversal** is a line, which intersects two or more parallel lines as shown below.

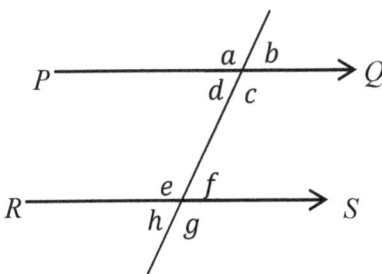

When a transversal intersects parallel lines as shown in the figure above, pairs of **corresponding angles**, **alternate angles** (sometimes referred to as Z-angles) and **adjacent angles** (or C angles) are formed as tabulated below.

Pairs of Alternate angles	Pairs of Corresponding angles	Pairs of Adjacent angles
$\hat{d}$ and $\hat{f}$	$\hat{a}$ and $\hat{e}$	$\hat{d}$ and $\hat{e}$
$\hat{c}$ and $\hat{e}$	$\hat{b}$ and $\hat{f}$	$\hat{c}$ and $\hat{f}$
	$\hat{c}$ and $\hat{g}$	
	$\hat{d}$ and $\hat{h}$	

Investigative Activity

1. In the figure above, measure the angles
   Record your result in the following table. (For a review of how to measures angles see Topic 10 of book 1)

Angle	$\hat{a}$	$\hat{b}$	$\hat{c}$	$\hat{d}$	$\hat{e}$	$\hat{f}$	$\hat{g}$	$\hat{h}$
Measured value(°)								

2. Which of the angles are equal?
3. List all the pair of angles whose sum is 180°.
4. Compare your result with the pair of angles in table (a) above.
5. Which of the pair of angles are vertically opposite?
6. Which of the pair of angles are alternate angles?
7. Which of the pair of angles are corresponding angles?
8. What conclusions do you draw?

The above investigation leads us into the following theorem.

Theorem

When a transversal intersects parallel lines,

1. Alternate angles are equal i.e. $\hat{d} = \hat{f}$ and $\hat{c} = \hat{e}$
2. Corresponding angles are equal
   i.e. $\hat{a} = \hat{e}$, $\hat{b} = \hat{f}$, $\hat{d} = \hat{h}$ and $\hat{c} = \hat{g}$
3. Adjacent angles are supplementary i.e.
   $\hat{d} + \hat{e} = 180°$ and $\hat{c} + \hat{f} = 180°$

Module 6, Topic 7: Angles

Example

In the figure below, *AB*, *DE* and *FH* are parallel. Find the value of *p*.

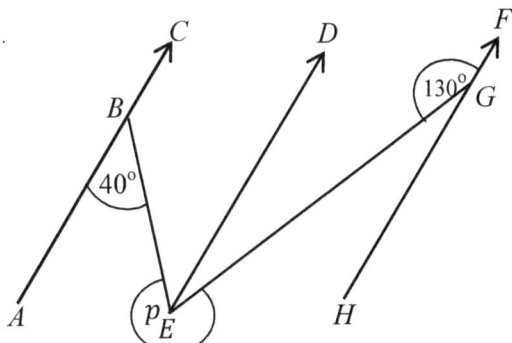

Solution

∠DEG + ∠EGF = 180° [Adjacent angles between ∥ lines]
∴ ∠DEG = 180° − 130 = 50°
∠ABE = ∠BED = 40° [Alternate angles between ∥ lines]
∴ ∠ABE = ∠BED + ∠DEG = 40° + 50° = 90°
∴ p = 360° − ∠BEG = 360° − 90° = 270°

Exercise 7:1

Find the value of the unknowns designated by the letter in figures (1) to (5) below.

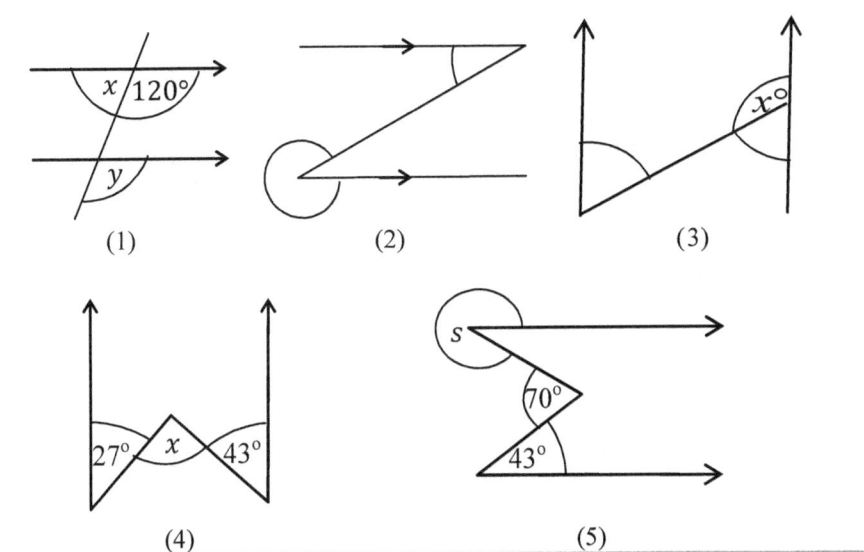

7.3 Interior and Exterior Angles of Polygons

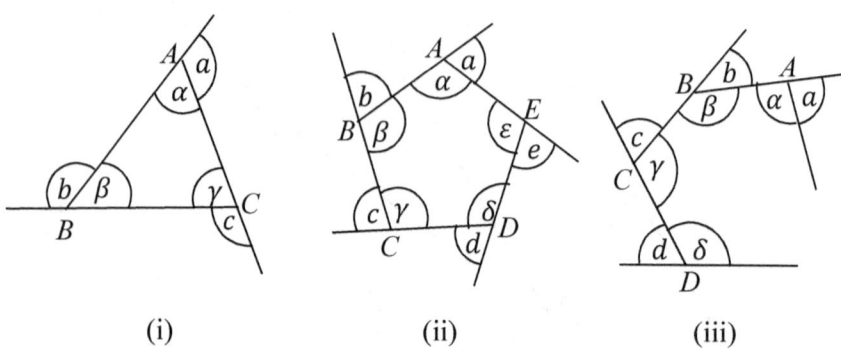

(i)   (ii)   (iii)

In the above figure, the angles α, β, γ, δ, ε, shown are called the **interior angles** of the polygons, because they are inside the polygons. On the other hand, the **exterior angles** of the polygons are the angles a, b, c, d, e formed outside each polygon when we produce AC, BA, CB, DC, ED respectively. Notice that each interior angle and the corresponding exterior angle are supplementary since they are angles on a straight line.

7.4 Sum of Angles of a Polygon

  Investigative Activity

1. Draw and label the vertices using the letters A, B, C...of a triangle, a quadrilateral, a pentagon, a hexagon and a heptagon
2. Draw lines from one particular vertex say A to all the other vertices.
3. Count the number of triangles into which the polygon has been divided and record your result in the table below.
   Use the following example which has been done for an octagon to help you.

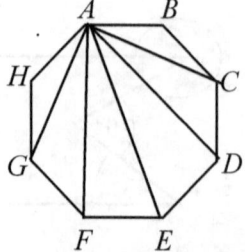

No of sides of polygon	No of triangles insides polygon	Sum of interior angles
3		
4		
5		
6		
7		
8	8	1080°

4. Study results in the table carefully and;
   (i) Predict the number of triangles inside a polygon with 9 sides.
   (ii) Predict the sum of all the interior angles in a convex polygon with 9 sides.
5. Write down an expression for the number of triangles insides a polygon with n sides.
6. Write down an expression the sum of interior angles of a polygon with n sides.
7. Deduce the sum of the exterior angles of a polygon.
8. Repeat instruction 1.
9. Draw lines from the vertex to the centre of each polygon as in the following figure.

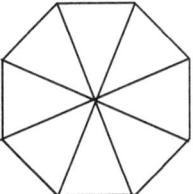

10. Count the number of triangles into which the polygon has been divided and record your result in the table below.
11. Study your results in the table carefully.
12. Predict the number of triangles inside a polygon with 9 sides.
13. What is the sum of the angles of all the triangles in a convex polygon with 9 sides?
14. What is the sum of the angles at the centre of the polygon?
15. Use (13) and (14) to find the sum of interior angles of a polygon with 9 sides.
16. What is the number of triangles inside a polygon with n sides?
17. What is the sum of the angles of all the triangles in a convex polygon with n sides?
18. Write down an expression for the sum of interior angles of a polygon with n sides.
19. Deduce the sum of exterior angles of a polygon.

No of sides of polygon	No of triangles insides polygon	Sum of interior angles
3		
4		
5		
6		
7		
8	8	1080°

The above investigative exercise, lead us into the polygon theorems that follow. The polygon theorems are actually extensions of Chasles' Theorem.

7.5 Polygon Theorems

In a convex polygon with n sides,

(1) The sum of the interior angles is $(n-2)180°$
(2) The sum of the exterior angles is 4 right angles (or 360°) no matter the value of n.

 Example

1. Each angle of a regular polygon is 170°. Find the number of sides of the polygon.

   Solution
   Each exterior angle = $180° - 170° = 10°$
   Since sum of exterior angles is equal to 360°, the number of sides must be $360° \div 10° = 36$.
   Therefore, the polygon has 36 sides.

2. One angle of a hexagon is 140° and 5 angles are equal. Find the value of each of the 5 angles.

   Solution
   The exterior angle corresponding to 140° is $180° - 140° = 40°$
   Therefore, the sum of the other 5 exterior angles is $360° - 40° = 320°$
   So the value of each of the 5 exterior angles $320° \div 5 = 64°$.
   Therefore, the corresponding angle is $180° - 64° = 116°$
   Hence, the value of each of the 5 angles is 116°.

Note!!

In solving problems on polygons it is often easier to use the theorem on sum of exterior angles of a polygon rather than that on the sum of interior angles, though both lead to the same answer.

 Exercise 7:2

1. Four angles of a hexagon are 130°, 160°, 112°, and 140°. If the remaining angles are equal, find the size of each.
2. The interior angle of a regular polygon is 140°. Find the number of sides of the polygon.
3. Each of the exterior angles of a regular polygon is 100° less than the interior angle. Calculate the size of the exterior angle.
4. How many sides, has a regular polygon whose interior angles are 108° each?
5. The sum of the interior angles of an n-sided convex polygon is double the sum of the exterior angles. Find the value of n.
6. The sum of the angles of a polygon is 1800°. Calculate the number of sides of the polygon
7. How many sides, has a convex polygon with each interior angle equal to 150°?
8. How many sides, has a convex polygon with each exterior angle equal to 20°?
9. Find the size of an interior angle of a regular ten-sided polygon.
10. Determine whether it is possible to have a regular convex polygon with interior angles; (a) 144°   (b) 140°   (c) 130°.
    If so state the number of sides of the polygon?
11. Determine whether it is possible to have a regular convex polygon with exterior angles; (a) 20°   (b) 16°   (c) 15°.
    If so state the number of sides of the polygon?
12. One exterior angle of a polygon is 54°, and other exterior angles are each 34°. Calculate the number of sides of the polygon.
13. Four of the angles of a pentagon are 72°, 100°, 120°, and 140°. Find the fifth angle.
14. AB, BC, CD are three consecutive sides of a regular polygon.
    If $\angle ACB = 15°$, calculate the number of sides of the polygon and $\angle ACD$.

Competency Base Mathematics for Secondary Schools Book 2

 Multiple Choice Exercise 7

1. By the number of sides the polygon in the figure below is:
   [A] an octagon   [B] a pentagon   [C] a quadrilateral   [D] a hexagon

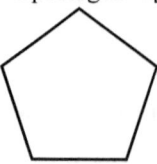

2. A seven sided plane figure is called:
   [A] an octagon   [B] a pentagon   [C] a hexagon   [D] a heptagon
3. A polygon with all its interior angles less than 180° is definitely:
   [A] a convex polygon   [B] a regular polygon
   [C] a re-entrant polygon   [D] a quadrilateral
4. The polygon which has the shape of the Cameroon flag is:
   [A] octagon   [B] pentagon   [C] hexagon   [D] quadrilateral
5. In the figure below, AB, CD and XY are straight lines intersecting at W. The value of $\angle CWX$ is:
   [A] 80°   [B] 100°   [C] 120°   [D] 140°

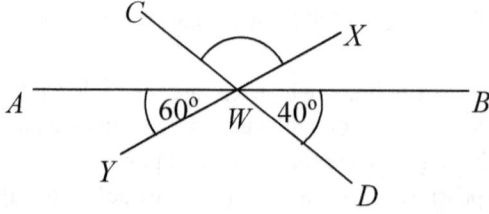

6. In the following figure, $PQ \parallel RS$ and the angles are shown. The size of x is:
   [A] 145°   [B] 150°   [C] 155°   [D] 165°

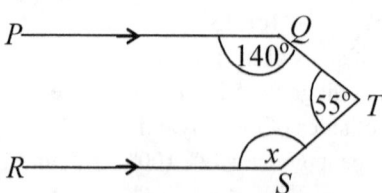

7. In the following figure, $AB \parallel CD$. The size of the angle marked x is:
   [A] 103°   [B] 93°   [C] 62°   [D] 52°

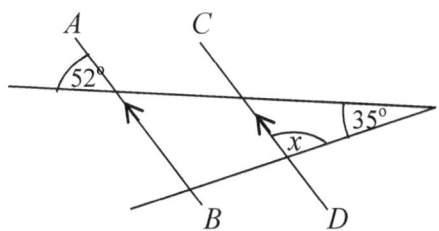

8. In the figure below, ∠PQU = 36°, ∠QRT = 29°, PQ∥RS and UQ∥RT. ∠PQR should be:
   [A] 94°   [B] 65°   [C] 61°   [D] 54°

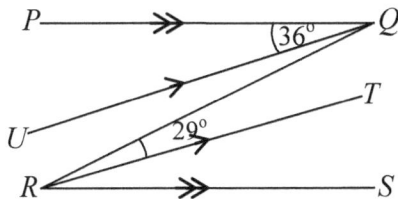

9. In the following figure, the value of x is:

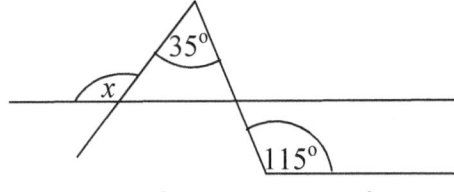

   [A] 35°   [B] 80°   [C] 100°   [D] 115°

10. In the figure below, PQ∥ST. The value of x is:
    [A] 82°   [B] 108°   [C] 124°   [D] 164°

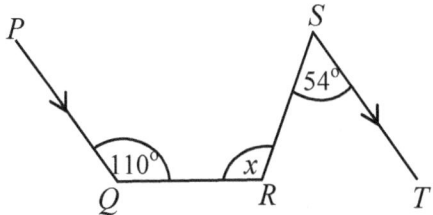

11. The number of sides in a regular polygon whose interior angle is 135° is:
    [A] 7   [B] 8   [C] 10   [D] 12

12. The number of sides in a regular polygon with each of its interior angles equal to 108° is:
    [A] 4   [B] 5   [C] 6   [D] 7

13. The number of sides in a regular polygon with each of its interior angles equal to 140° is:
    [A] 7   [B] 8   [C] 9   [D] 10

14. The number of sides in a regular polygon with each of its interior angles 120° is:
    [A] 4   [B] 6   [C] 7   [D] 8
15. In the figure below, y is equal to:
    [A] 80°   [B] 70°   [C] 40°   [D] 100°

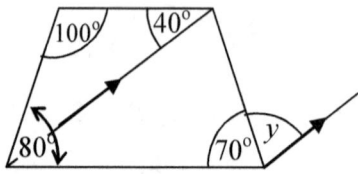

16. In the figure below, QRS is a straight line, QP∥RT, $\angle PQR = 56°$ $\angle QPR = 84°$, $\angle TRS = x°$. The value of x is:
    [A] 28°   [B] 40°   [C] 44°   [D] 84°

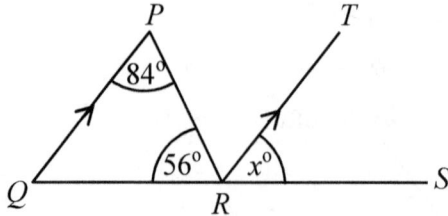

17. The number of sides in a regular polygon with one interior angle 160° is:
    [A] 10   [B] 36   [C] 18   [D] 20
18. In following figure, PQ is parallel to RST, $\angle PQR = 35°$ and $|RS| = |SQ| = |TQ|$. The size of $\angle STQ$ is:
    [A] 35°   [B] 40°   [C] 70°   [D] 110°

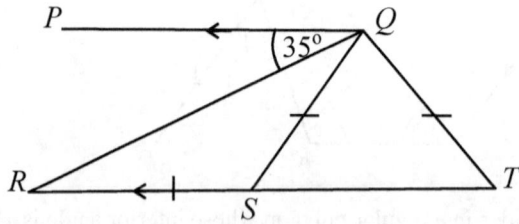

19. In the figure below, WXYZ is a rhombus and $\angle WYZ = 20°$. The value of angle XZY is:
    [A] 20°   [B] 30°   [C] 60°   [D] 70°

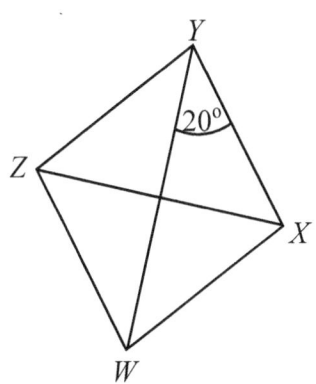

20. The number sides in a regular polygon with one interior angle 160° is:
    [A] 10   [B] 36   [C] 18   [D] 20
21. The diagram below which shows an impossible situation is:

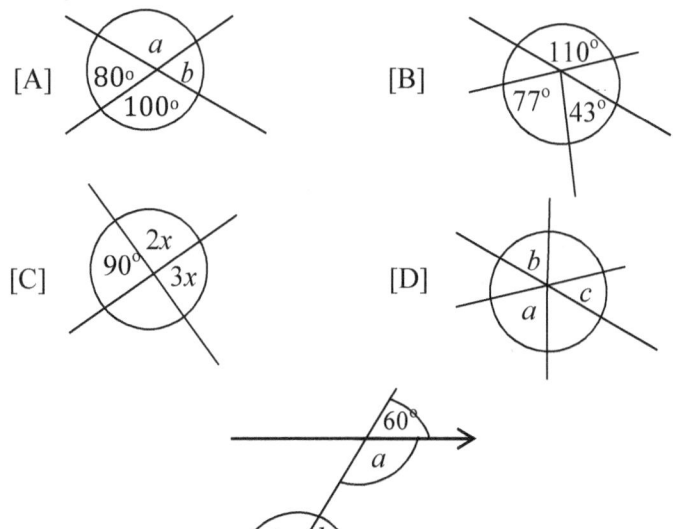

22. The angle *a*, shown in the figure above is equal to:
    [A] 60°   [B] 120°   [C] 30°   [D] 90°
23. The angle *b* shown in figure (b) above is equal to:
    [A] 60°   [B] 120°   [C] 30°   [D] 90°
24. The angle *c* shown in figure (b) above is equal to:
    [A] 60°   [B] 120°   [C] 30°   [D] 90°

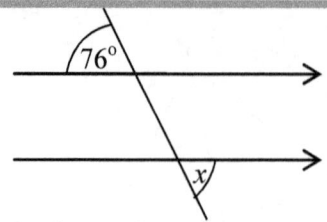

25. The value of x in the figure above is:
    [A] 76°   [B] 104°   [C] 14°   [D] 36°
26. In the figure below, LK ∥ PQ. ∠KLM = 241° and ∠QPM = 89°. The value of ∠LMP is:
    [A] 61°   [B] 30°   [C] 119°   [D] 150°

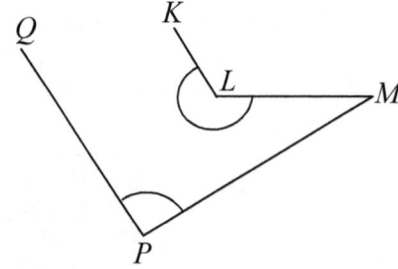

27. In the figure below, ML ∥ PQ and NP ∥ QR. Given that ∠LMN = 40° and ∠MNP = 55° then ∠RQP equals:
    [A] 15°   [B] 25°   [C] 35°   [D] 40°

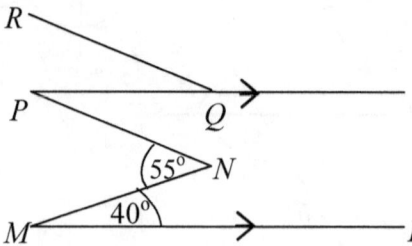

Topic 8

TRIANGLES

Objectives

At the end of this topic, the learner should be able to:

1. Pythagorean triples.
2. Angles in a triangle.
3. Determine the sum of the interior angles of a triangle.
4. Exterior angle and corresponding opposite interior angles
5. Congruent triangles
6. State the Pythagoras theorem and its converse.
7. Solve a right angled triangle using Pythagorean triples.

8.1 Standard Notation for Triangles

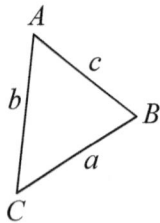

In labeling a triangle, uppercase or capital letters are used to label the vertices (corners) as shown in the figure above. The side opposite a vertex is labeled using its corresponding lowercase or small letter. For instance the side opposite the vertex A is labeled using the letter a. The triangle above is referred to as triangle ABC. The order of the letters does not matter.

8.2 Interior and exterior angles of Triangles

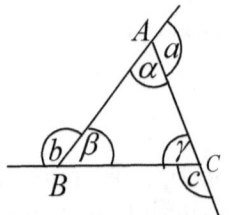

In the figure above, the angles α, β and γ are called the **interior angles** of the triangle, because they are inside the triangle. On the other hand, the angles a, b, c, formed outside the triangle when the sides AC, BA, CB, respectively are produced are called the **exterior angles** of the triangle. Notice that each interior angle and the corresponding exterior angle are supplementary (sum up to $180°$) since they are angles on a straight line.

 Investigative Activity

1. Draw a triangle ABC (any size) and produce AC to D as shown in figure (a) below.
2. Cut out the vertices A and B of the triangle ABC and arranged at C as in figure (b) below.
4. Do they fit exactly onto the angle θ the exterior angle of the triangle?
5. Do the three angles α, β and γ lie and fit exactly on the straight line AD?
6. What is the numerical value of $\alpha + \beta + \gamma$?
7. What conclusion do you draw?

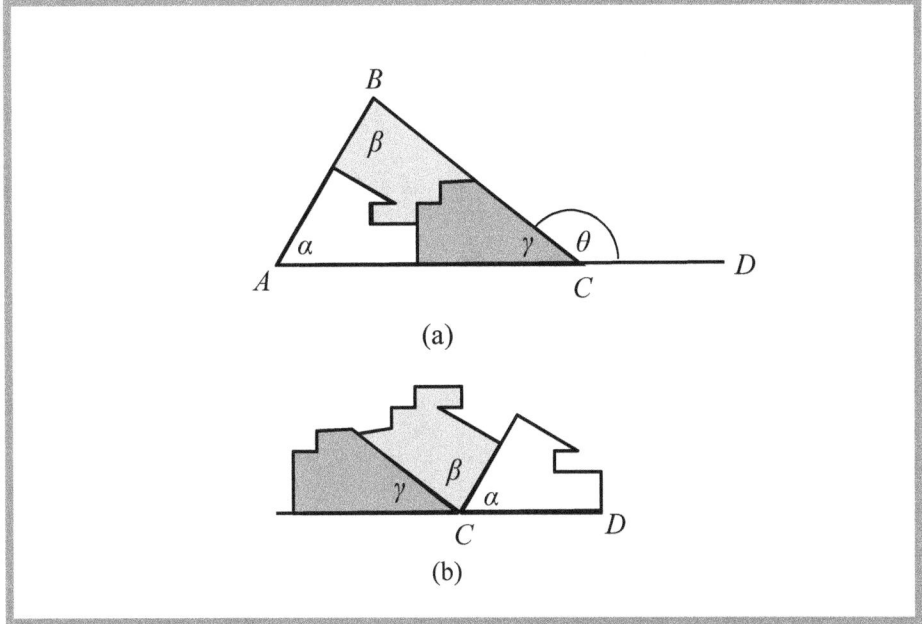

(a)

(b)

The above investigation leads us to the following theorem called Chasles theorem

.

8.3 Chasles' Theorem

1. The exterior angle of a triangle is equal to the sum of the two interior opposite angles. i.e. $\theta = \alpha + \beta$
2. The sum of the interior angles of a triangle is two right angles (or 180°). i.e. $\alpha + \beta + \gamma = 180°$.

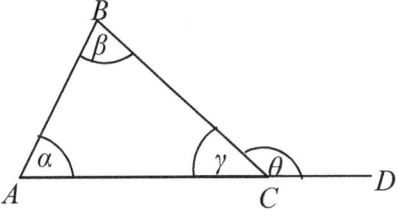

Exercise 8:1

Find the value of the lettered angles in figure (a) to (i) below.

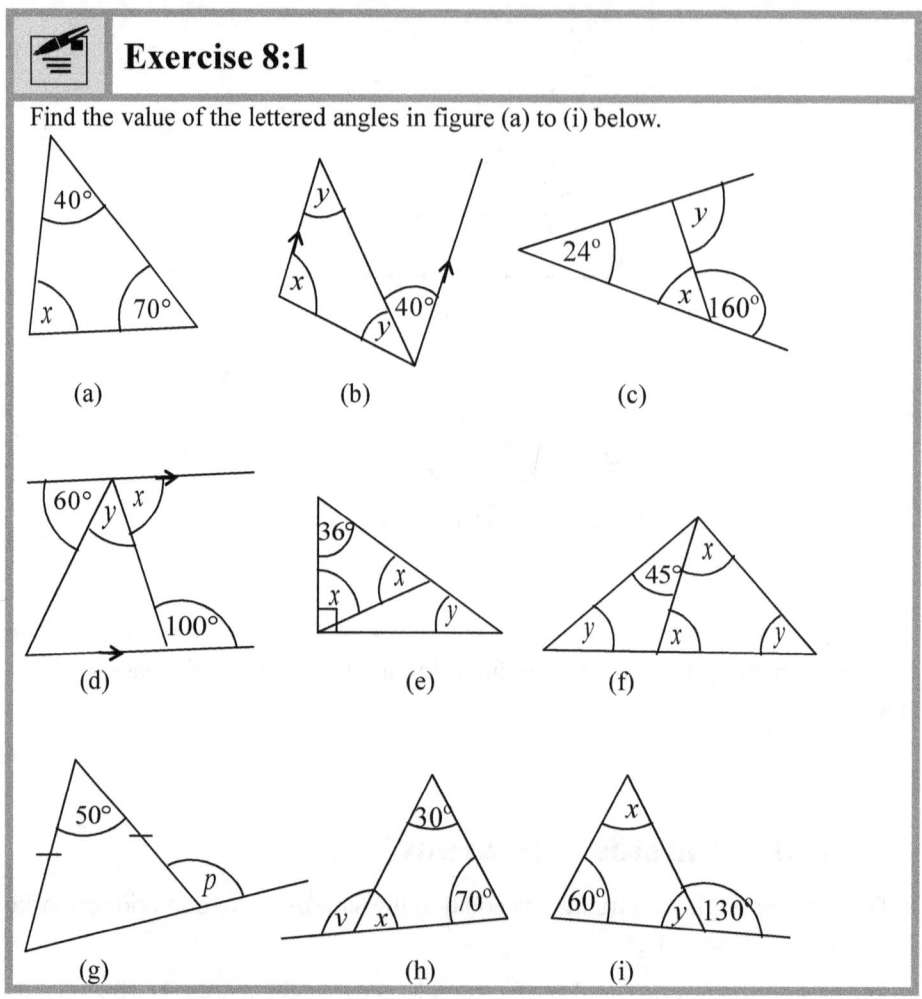

8.4 The Right-Angled Triangle

A **right-angled triangle** is a triangle with one of its angles equal to 90°. The other two angles are each less than 90°. An angle, which is equal to 90°, is called a **right angle**. A right angle is usually marked using a small square as shown in Figure 18:14. An angle, which is greater than or equal to 90° but less than 180° is called an **acute angle**. In a right-angled triangle, the longest side AC is called the **hypotenuse** and is the side opposite to the right angle. The two shorter sides AB and BC are next to the right angle and are called the **arms** or the **legs** or the **limbs** of the right-angled triangle.

Module 6, Topic 8: Triangles

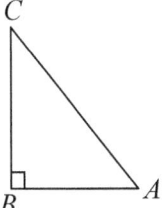

In the figure above, the sides *a* and *c* can be reckoned in the following ways;

Side *a* is opposite to angle *BAC* or *a* is adjacent or next to angle *ACB*.

Side *c* is opposite to angle *ACB* or *c* is adjacent or next to angle *BAC*.

 Investigative Activity

The triangle below shows a right-angled triangle *ABC* with *AB* = 3 squares and *BC* = 4 squares.

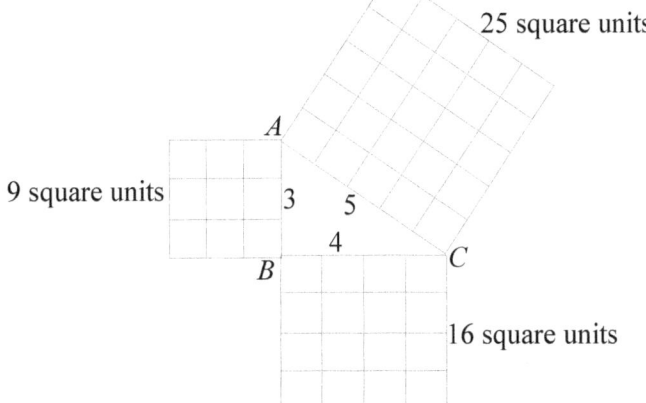

1. Count the number of squares on each arm of the triangle and record your result in the table below.
2. Also count the number of squares on the hypotenuse and record your result in the table below.
3. On a square paper draw a right-angled triangle *ABC* with *AB* = 5 squares and *BC* = 12 squares.
4. Cut out part of the square paper along one of the lines and place on the hypotenuse *AC* of the triangle as in the right-angled triangle above.
5. What is the length of the hypotenuse in terms of the number of squares?
6. Draw the large squares on each side of the triangle.
7. Count the number of squares on each arm of the triangle and record your result in table below.
8. Also count the number of squares on the hypotenuse and record your result in table below.

9. Draw other right-angled triangles with sides of different lengths, and repeat the instructions in numbers 4 to 8.
10. Study your table carefully.
11. What conclusion do you draw?

Number of squares on side			Sum of squares on AB and BC = a + c
AB = c	BC = a	AC = b	

The above investigation leads us to the following theorem called the Pythagoras theorem.

8.5 The Pythagoras Theorem

The Pythagoras theorem states that in a right-angled triangle, the square on the hypotenuse is equal to the sum of the squares on the other two sides.

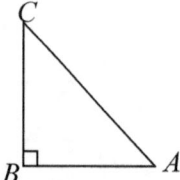

Thus for the right-angled triangle above,

$$AC^2 = AB^2 + BC^2$$
$$\Rightarrow AC = \sqrt{AB^2 + BC^2}$$
$$\Rightarrow AB = \sqrt{AC^2 - BC^2}$$
$$\Rightarrow BC = \sqrt{AC^2 - AB^2}$$

Conversely, any triangle which is such that the sum of the squares on two sides is equal to the square on the third side is a right-angled triangle.

 Example

Find the length AB of the longest side of the right-angled triangle ABC with the other sides, AC and BC given as 3 cm and 4 cm respectively.

Solution

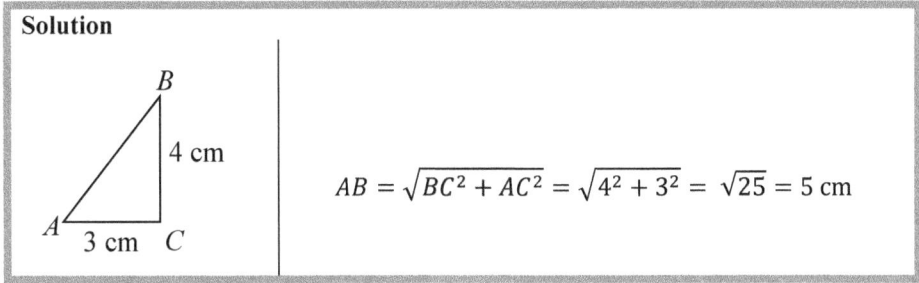

$$AB = \sqrt{BC^2 + AC^2} = \sqrt{4^2 + 3^2} = \sqrt{25} = 5 \text{ cm}$$

Test Rule for a Right-Angled Triangle

If in a triangle whose sides are known, the sum of the squares on two of the sides is equal to the square on the other side, the triangle is a right-angle triangle. Otherwise the triangle cannot be a right-angled triangle.

Pythagorean Triples

Any three whole numbers, which can form sides of a right-angled triangle, are called Pythagorean triples. Examples of Pythagorean triples are 3,4,5; 6,8,10; 5,12,13; 8,15,17; 7,24,25 etc.

Multiples of Pythagorean Triples

Consider the Pythagorean triple 3, 4, 5. The table below shows the result of multiplying each of the numbers in the triplet by the same factor n.

n	AB	BC	AC	AB^2	BC^2	AC^2	$AB^2 + BC^2$
1	3	4	5	9	16	25	25
2	6	8	10	36	64	100	100
3	9	12	15	81	144	225	225
4	12	16	20	144	256	400	400
5	15	20	25	225	400	625	625

From the table, we can see that multiples of the Pythagorean triple 3, 4, 5 are also Pythagorean triples.

Generally,

Multiples of Pythagorean triples are also Pythagorean triples.

Exercise 8:2

1. Find the unknown in each of the following triangles.

   (a) 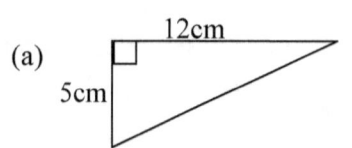   (b)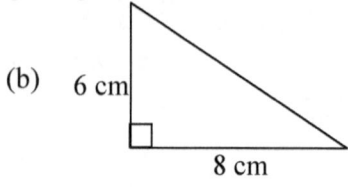

   (c) 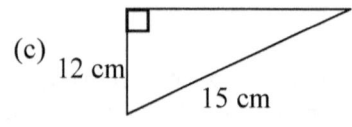   (d)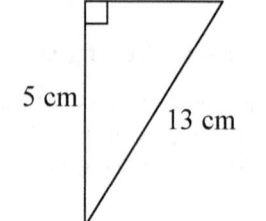

   (e) 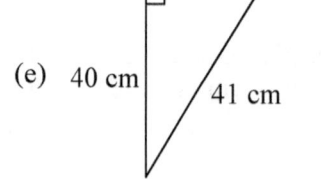   (f)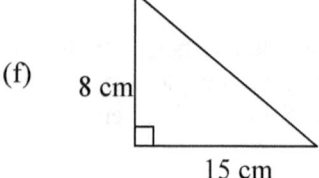

   (g) 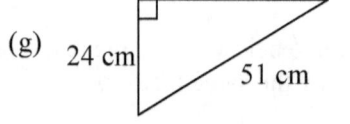   (h)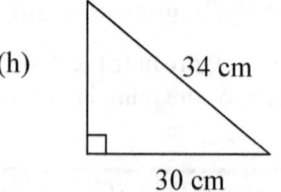

   (i) 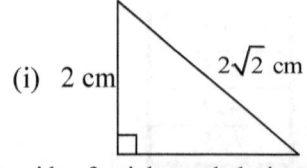   (j)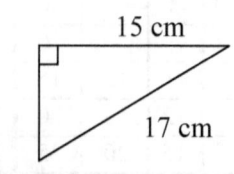

2. One side of a right-angled triangle is 3 cm. If the hypotenuse is 5 cm, find the length of the other side.
3. Find the diagonal of a rectangle whose sides are 40 cm by 9 cm long.
4. The diagonal of a rectangle is 30 cm and one of its sides is 24 cm; find the other side of the rectangle.
5. In a right-angled triangle whose hypotenuse is 20 cm, the ratio of the two arms is 3: 4. Find each arm.
6. Determine which of the following triplets Pythagorean triples are.
   (a) 17,15,8   (b) 6,9,11   (c) 40,41,9   (d) 3,2,5   (e) 5,12,13
   (f) 7,9,12   (g) 14,17,20   (h) 6,8,10   (i) 3,6,8

Module 6, Topic 8: Triangles

8.6 Congruent Triangles

 Investigative Activity

1. Draw a triangle on a cardboard.
2. Cut it out with a razor blade or scissors.
3. Place it on a square paper and retrace it carefully with a pencil. Label this triangle *A* as on the grid below.

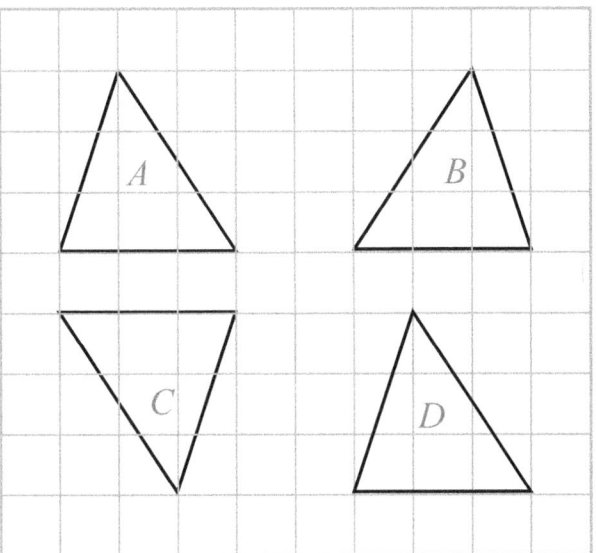

4. Turn the cardboard triangle over, then place it at another position and retrace it again. Label it *B*.
5. Rotate the cardboard triangle, then place it on another position and retrace it again. Label it *C*.
6. Place the cardboard triangle at *A* and arrange it to fit, then push it on another position and retrace it again. Label it *D*.
7. What can you say about the size and shape of the triangles *A*, *B*, *C* and *D*?
8. What can you say about the matching sides of the triangles *A*, *B*, *C* and *D*?
9. What can you say about the matching angles of the triangles *A*, *B*, *C* and *D*?

The triangles A, B, C and D which you used the cardboard triangle to retrace are congruent triangles. Congruent figures are figures that have the same shape and the same size. Congruent plane figures are figures, which can fit on each other. The statement '*A* is congruent to *B*' is written *A*≡*B*.

8.7 Conditions for Triangles to be Congruent

1. If two triangles are such that the three sides of one are equal to the three sides of the other as in the figure below, then they are congruent by **side-side-side** abbreviated **SSS**.

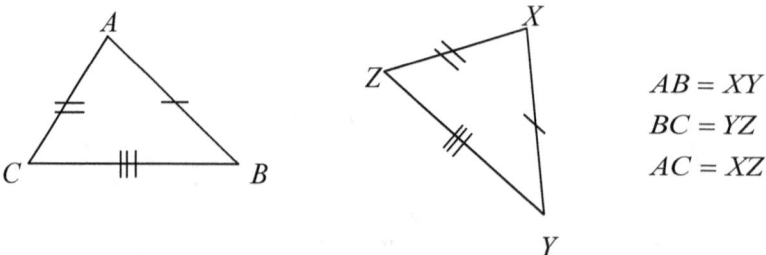

$AB = XY$
$BC = YZ$
$AC = XZ$

2. If two triangles are such that two sides of one are equal to two sides of the other and the included angles are equal as in the figure below, then they are congruent by **side -included angle-side** abbreviated **SAS**.

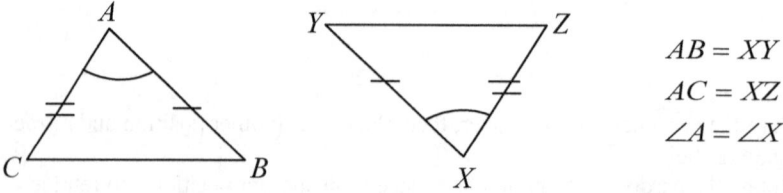

$AB = XY$
$AC = XZ$
$\angle A = \angle X$

3. If two triangles are such that two angles of one are equal to two angles of the other and one side of another is equal to one side of the other as in the figure below, they are congruent by **angle-side-angle** abbreviated **ASA**.

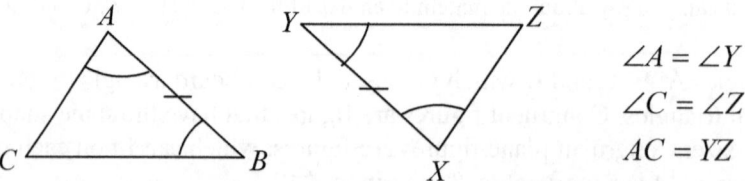

$\angle A = \angle Y$
$\angle C = \angle Z$
$AC = YZ$

4. If two right-angled triangles have equal hypotenuse and one arm of another is equal to one arm of the other as in the figure below, then they are congruent by **right angle-hypotenuse- side** abbreviated **RHS**.

Module 6, Topic 8: Triangles

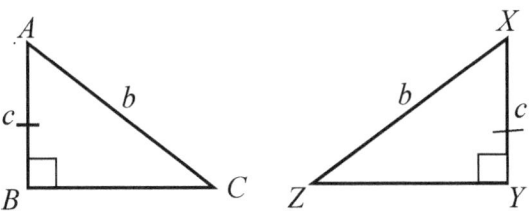

$\angle B = \angle Y = 90°$

$AC = XZ = b$

$AB = XY = c$

 Example

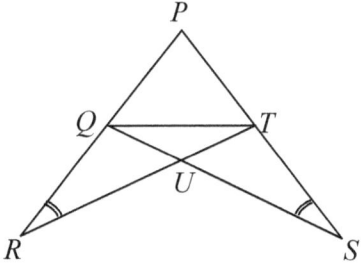

In the figure below, $PR = PS$ and Q and T are the mid-points of PR and PS respectively. State the number of pairs of congruent triangles in the figure and name them, stating the matching sides for each pair of congruent triangles.

Solution
There are three pairs of congruent triangles as follows.
$\triangle PTR \equiv \triangle PQS$. $PT = PQ, PR = PS$ and $RT = SQ$.
$\triangle QTR \equiv \triangle TQS$. QT is common, $QR = TS$ and $RT = SQ$.
$\triangle QUR \equiv \triangle TUS$. $QU = TU, QR = TS$ and $RU = SU$.

Exercise 8:3

In problems 1 to 7, State the number of pairs of congruent triangles in the figure and name them, stating the matching sides for each pair of congruent triangles.

1. In the figure below PQ is parallel to SR.

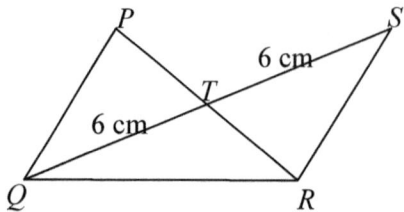

2. In the figure below triangle ABE is isosceles $BC = DE = 3$ cm.

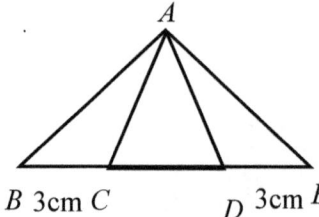

3. In the figure below, O is such that $OY = OZ$ and $\angle YOZ = \angle ZOX = \angle XOY = 120°$.

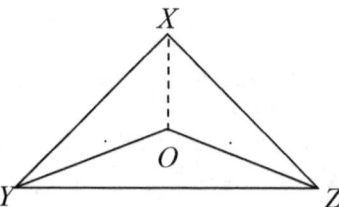

4. In the figure below, $OM = NL$, $\angle OMP = \angle LNQ$ and $\angle POM = \angle QLN$.

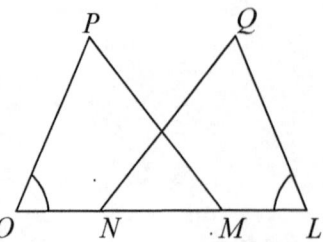

5. In the figure below, $PQRS$ is a parallelogram.

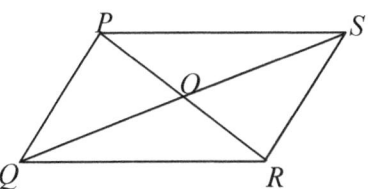

6. In the figure below, $PT = TR$ and PQ is parallel to SR.

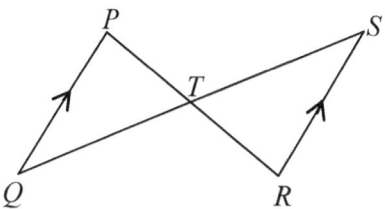

7. In figure (a) below, $XO = WO$ and XW is parallel to YZ.

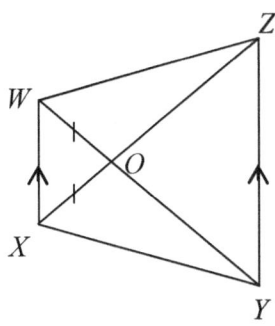
(a)

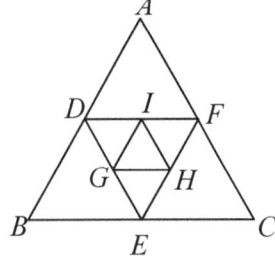
(b)

8. In figure (b) above, triangle ABC is isosceles and D, E and F are midpoints of AB, BC and AC respectively. How many sets of congruent triangles are there in the figure? List each of these sets.

Multiple Choice Exercise 8

1. The largest angle of a triangle:
   [A] Must always be an acute angle.   [B] Can sometimes be an acute angle.
   [C] Can never be a right-angle.   [D] Must always be an obtuse angle.
2. In figure below, the size of angle ACB is:

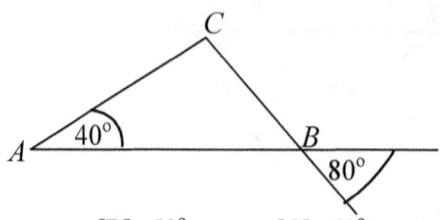

[A] 40°   [B] 50°   [C] 60°   [D] 80°

3. In the figure below, $|PQ| = |PR| = |RS|$ and $\angle RPS = 32°$. The value of $\angle QPR$ is:

   [A] 64°   [B] 52°   [C] 32°   [D] 26°

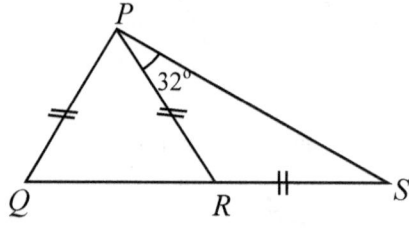

4. In the following figure, ABC is a triangle, BC is produced to D, $|AB| = |AC|$, $\angle BAC = 50°$. The value of $\angle ACD$ is:

   [A] 115°   [B] 65°   [C] 60°   [D] 50°

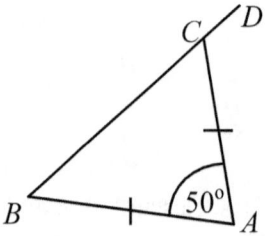

5. The value of angle t in the figure below is:

   [A] 115°   [B] 120°   [C] 125°   [D] 145°

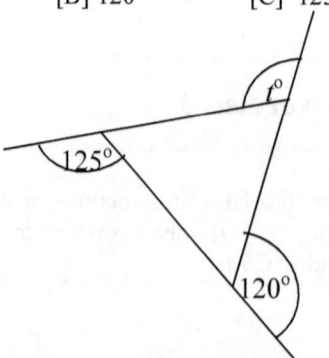

6. A Pythagorean triple among the following is:

[A] 6, 9, 11   [B] 7, 9, 12   [C] 8, 15, 17   [D] 14, 17, 20

7. The triplet which does not represent the lengths of the sides of a right-angled triangle is:
 [A] 6, 8, 10   [B] 5, 12, 10   [C] 8, 15, 17   [D] 7, 23, 24

8. A triangle has sides 8 cm, 15 cm and 17 cm. Therefore, the best name for it is:
 [A] an equilateral triangle.   [B] an obtuse triangle.
 [C] a right-angled triangle.   [D] an isosceles triangle.

9. The area of a square is equal to that of a triangle of base 9 cm and altitude 32 cm. The length of the side of the square must be:
 [A] 6 cm   [B] 6.2 cm   [C] 12 cm   [D] 2.2 cm

10. A square has a diagonal of 10 cm. the length of a side of the square is:
 [A] $\sqrt{10}$ cm   [B] $\sqrt{50}$ cm   [C] 10 cm   [D] 5 cm

11. The pair of triangles below which is definitely congruent is:

 [A]   [B]

 [C]   [D]

12. The pair of triangles below which is definitely congruent is:

 [A]   [B]

 [C]   [D]

13. The triangles PQR and DEF are equiangular but not congruent. The triangles DEF and XYZ are congruent. It follows that:
 [A] $\triangle PQR$ and $\triangle DEF$ are equal in area
 [B] $\triangle PQR$ and $\triangle XYZ$ are congruent
 [C] $\triangle PQR$ and $\triangle XYZ$ are equal in area
 [D] $\triangle PQR$ and $\triangle XYZ$ are similar

14. The pairs of triangles PQR, XYZ are congruent if:
 [A] $XY = PQ, XZ = QR, \angle X = \angle Q$
 [B] $XZ = QR, YZ = PR, \angle Y = \angle P$
 [C] $\angle Y = \angle P, \angle Z = \angle Q, XZ = PQ$
 [D] $\angle Z = \angle P, \angle Y = \angle Q, XY = PR$

Module 7

Solid Figures

Family of Situations

Module 7 is an extension of module 3 and at the end of the module; the student is expected to acquire many more competencies within the **families of situations** *'Usage of Technical Objects in everyday life'*.

Categories of Action

The categories of action for module 7 include:
1. Production of commodities or provision for daily consumption,
2. Production of parts for industrial use,
3. Production of materials for work of arts and construction.

Credit

The module is expected to be covered within 4 weeks teaching 4 hours per week (or within 15 to 16 hours).

Topic 9

PRISMS AND CYLINDERS

Objectives

At the end of this topic, the learner should be able to:

1. Observe and describe a prism
2. Recognize and identify a right prisms
3. Identify the apex, lateral surface, lateral edge, altitude of a prism.
4. Make sketches of prisms.
5. Draw and make net of prisms
6. Make models of prisms from nets.
7. Use the various parts of the net to establish the original figure.
8. Calculate total surface area and volume of prisms.
9. Calculate total surface area and volume of cubes.
10. Calculate total surface area and volume of cylinders.

9.1 Observation and Description of Prisms

A prism is a solid with uniform polygonal cross-sectional area. This definition means that the lateral surfaces of prisms are made up of rectangles.

Prisms are named from the nature of their cross-sections as shown in the following table and figure (i) and (ii) below.

Nature of cross-section	Name of Prism	Diagram of Prism
Triangle	Triangular Prism	
Square	Square Prism or cube	
Rectangle	Rectangular Prism or cuboid	
Pentagon	Pentagonal Prism	
Hexagon	Hexagonal Prism	

A **right-angled triangular prism** (figure (i) below) is a prism whose cross-section is a right-angled triangle.

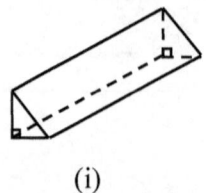 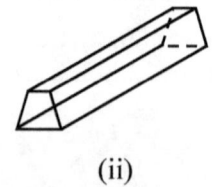

(i)   (ii)

Right-angled triangular prism   Trapezoidal prism

A **trapezoidal prism** (figure (ii) above) is a prism whose cross-section is a trapezium.

A prism is called a **right prism** (figure (i) below) if the lateral surfaces are made up of rectangles; otherwise, it is called an **oblique prism** (figure (ii) below).

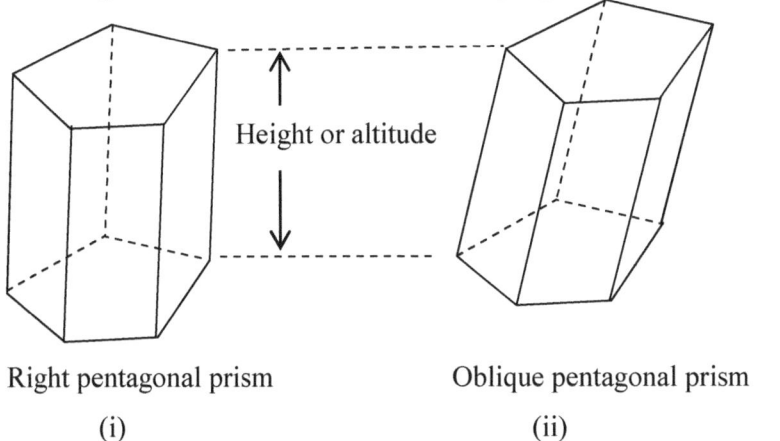

Right pentagonal prism
(i)

Oblique pentagonal prism
(ii)

9.2 Edges, Vertices, Faces of Solid Figures

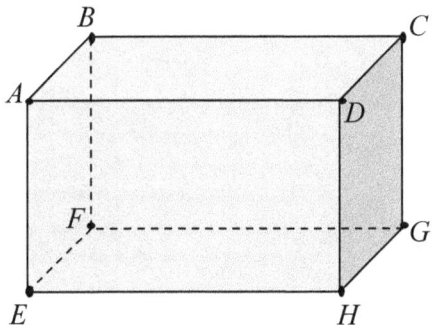

The plane sides of a solid figure such as *ABCD* and *ADEH* are called **faces.** The sharp lines such as *AD* where the plane faces of a solid figure meet are called **edges**. The pointed corners such as *A, B, C* and *D* of a plane or solid figure are called **vertices**. In the figure above, the cuboid is drawn in perspective so that the three hidden faces, the three hidden edges represented by dotted lines and the one hidden vertex can be seen. In this figure, the vertices are represented by dots, the straight lines are the edges and the plane rectangular sides are the faces. From the figure, it is clear that there are 6 faces, 12 edges and 8 vertices.

9.3 Nets of Prisms

Consider the shapes in figure below called **nets of polygons**. If each is cut and folded along, the dotted lines so that the ends meet, (i) will form a triangular prism. Which solid figure will (ii) form if folded along the dotted lines?

Competency Base Mathematics for Secondary Schools Book 2

(i) (ii)

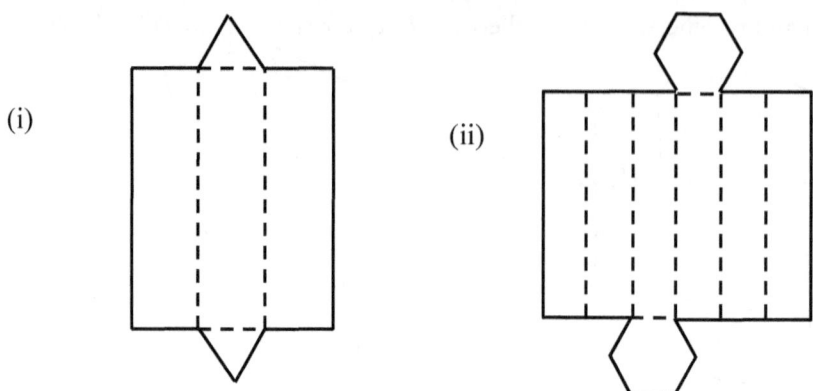

 Exercise 9:1

1. How many faces bound the following?
   (i) A cube   (ii) A trapezoidal prism
2. Copy and complete the following table.

Name of Prism	Number of		
	Faces	Edges	Vertices
Triangular prism			
Cube			
Cuboid			
Pentagonal prism			
Hexagonal prism			

9.4 Surface Area and Volume of Prisms

A right prism has a pair of parallel and congruent polygonal base.

Consider a prism of length l, cross-sectional area A and cross-sectional perimeter p.

The surface area S of the prism is given by

$S = 2A + pl$ ....①

The volume V of the prism is given by

$V = Al$ ....②

Module 7, Topic 9: Prisms and Cylinders

Example

1. A right-angled triangular prism has arms 4 cm and 3 cm and length 10 cm. Calculate (a) its surface area. (b) its volume.

   Solution

   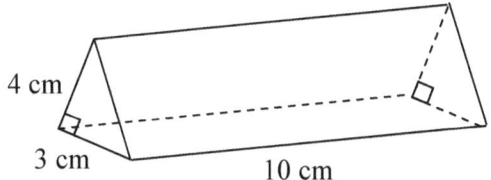

   (a) $S = 2A + pl$
   Hypotenuse of triangle $= \sqrt{3^2 + 4^2} = 5$ cm
   $\Rightarrow p = 3 + 4 + 5 = 12$ cm
   $A = \frac{1}{2}bh = \frac{1}{2}(4)(3) = 6$ cm^2
   Therefore, $S = 2(6 \text{ cm}^2) + (12 \text{ cm})(10 \text{ cm}) = 132$ cm^2

   (b) $V = Al = 6 \text{ cm}^2 (10 \text{ cm}) = 60$ cm^3

2. The sides of a rectangular block (cuboid) are 5 cm, 11 cm and 40 cm. Calculate (a) its surface area. (b) its volume.

   Solution

   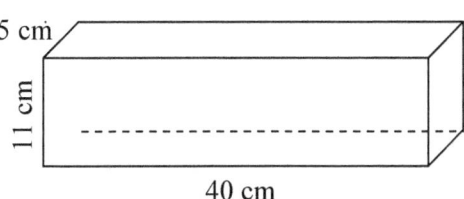

   (a) $S = 2A + pl = 2(11)(5) + 2(5 + 11)(40)$
   $= 110 + 1280 = 1390$ cm^2

   (b) $A = (5)(11) = 55$ cm^3
   $V = Al = (55 \text{ cm}^2)(40 \text{ cm}) = 2200$ cm^3

9.5 Surface Area and Volume of Cubes

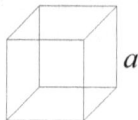

A cube is a special prism with the length, width and height equal.
Applying equation ① and ② to a cube of side a,
$S = 2A + pl$, $l = a, A = a^2$ and $p = 4a$
$S = 2a^2 + 4(a)(a) = 6a^2$
Therefore for a cube, $S = 6a^2$ and $V = a^3$

 Example

Example 21:3
A cube has side 9 cm. Find (a) its surface area. (b) its volume.

Solution
(a) $A = 6a^2 = 6(9)^2 = 486$ cm² (b) $V = a^3 = 9^3 = 729$ cm³

 Exercise 9:2

1. A box is 35 cm long, 20 cm wide and 12 cm high. Calculate the volume of the box.
2. What volume of sand will fill a room of length 15 m, breadth 12 m and height 8 m?
3. A box has a volume of 140,000 cm³. If its breadth is 50 cm, and its length is 70 cm, find its height.
4. The volume of a rectangular tank is 520 m³. Find the length of the tank given that the width and the height are 8 m and 5 m respectively.
5. Find the level that 270 m³ of water will rise in a rectangular tank which is 6 m by 4.5 m.
6. Find the volume of rectangular block 0.5 m long and cross-sectional dimensions of 3 cm by 20 cm.
7. Calculate the number of litres of water, which a rectangular tank with dimensions 8m by 6m by 3 m will hold.

Module 7, Topic 9: Prisms and Cylinders

8. How many litres of water will fill a rectangular tank of 500 cm by 100 cm by 200 cm?
9. Find the capacity in litres of a rectangular tank with dimensions 3 m by 4m by 5 m.
10. A cube measures 2 cm. Calculate
    (a) The volume of the cube
    (b) The total surface area of the cube
11. Calculate the number of blocks each of dimension 2.5 cm by 5 cm by 7.5 cm that can be stored in a box of dimensions 1 dm by 3 dm by 6 dm.
12. The end of a prism is a right-angled triangle, with the sides containing the right angle 8 cm by 12 cm. If the prism is 20 cm long, what is,
    (a) its volume?   (b) its surface area?
13. A prism is 10 cm long. Its cross-section is an equilateral triangle of side 8cm calculated
    (a) The volume of the prism   (b) The surface area of the prism.

9.6 Surface Area and Volume of Cylinders

A cylinder is similar to a prism because they both have uniform cross-sections. The cross-section of a cylinder is circular. When a cylinder is closed at both ends, it is described as a **solid cylinder**.

Applying equation ① and ② to a solid cylinder with radius r and length l,

$$S = 2A + pl \text{ and } V = Al$$

The perimeter p here is the circumference and the area A is that of the circular cross-section.

$$\Rightarrow p = 2\pi r \text{ and } A = \pi r^2$$
$$\Rightarrow S = 2\pi r^2 + 2\pi rl$$

Therefore for a right solid cylinder,

$$S = 2\pi r(r+l) \text{ and } V = \pi r^2 l$$

Example

A right-circular solid cylinder has a height of 30 cm and a radius of 7 cm. Calculate (a) its surface area (b) its volume

Solution

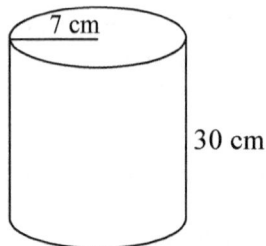

(a) $S = 2\pi r(r+l) = 2\left(\dfrac{22}{7}\right)(7)(7+30) = 1{,}628$ cm^2

(b) $V = \pi r^2 l = \left(\dfrac{22}{7}\right)(7)^2(30) = 4620$ cm^3

 Exercise 9:3

In this exercise, take $\pi = \dfrac{22}{7}$ where necessary.

1. The height of a cylinder with radius 35 cm is 21 cm. Find the curved surface area of the cylinder.
2. Find the area of the curved surface of a cylinder whose radius is 7 cm and whose height is 5 cm.
3. Find the height of a cylinder with radius 7 cm if its volume is 770 cm^3.
4. The following table shows the dimensions of a cylinder. Complete the table.

(a)	Base radius	5 cm	10 cm
(b)	Height	20 cm	30 cm
(c)	Surface area		
(d)	Volume		

5. A cylindrical tank 3.5 m in diameter contains water to a depth of 4 m. Find the total area of the wetted surface of the tank.
6. A cuboid of sides 2 cm by 4 cm by 11 cm is full of water. If this water is poured

Module 7, Topic 9: Prisms and Cylinders

into a cylindrical jar of diameter 8 cm, find the depth of the water.
7. Calculate the volume of a cylindrical container with diameter 14 cm and height 7 cm.
8. The volume of a cylinder is 396 cm³. Calculate the radius of the cylinder given that the height is 14 cm.
9. A drum is 20 cm in diameter and 70 cm tall. Calculate the capacity of the drum.
10. How many litres of liquid can a cylindrical can 14 cm in diameter and 20 cm high contain?
11. 90 litres of water is poured into a cylindrical bucket, which is 30 cm in diameter. Find the depth of water in the bucket.
12. The base of a pyramid is a square. Each face is made up of isosceles triangles with base 12 cm and height 16 cm. Calculate
    (a) its surface area   (b) its volume

Multiple Choice Exercise 9

In this exercises, where necessary, take $\pi = \frac{22}{7}$.

1. The shape of each side of a cuboid is:
   [A] A triangle   [B] A trapezium
   [C] A circle   [D] A rectangle
2. The number of vertices in a cuboid is:
   [A] 4   [B] 6   [C] 8   [D] 12
3. The number of faces in a cuboid is:
   [A] 4   [B] 6   [C] 8   [D] 12
4. The number of edges in a cuboid is:
   [A] 12   [B] 8   [C] 6   [D] 4
5. The total surface area of a cube of edge 3 cm is:
   [A] 27 cm²   [B] 27 cm³   [C] 54 cm²   [D] 36 cm²
6. The sides of two cubes are in the ratio 2:5. The ratio of their volumes is:
   [A] 4:5   [B] 8:15   [C] 6:125   [D] 8:125
7. A rectangular tank 2.25 m long and 1.6 m wide contains 2800 litres of water. Correct to the nearest cm, the depth of water in the tank is:
   [A] 76 cm   [B] 78 cm   [C] 770 cm   [D] 780 cm
8. A cylindrical container closed at both ends, has a radius of 7 cm and a height 5 cm. The total surface area of the container is:
   [A] 154 cm²   [B] 220 cm²   [C] 528 cm²   [D] 770 cm²
9. A cylindrical container closed at both ends, has a radius of 7 cm and a height 5 cm. The volume of the container is:
   [A] 154 cm³   [B] 220 cm³   [C] 528 cm³   [D] 770 cm³
10. The curved surface area of a cylindrical tin is 704 cm². The height when the radius is 8 cm is:
    [A] 3.5 cm   [B] 7 cm   [C] 14 cm   [D] 6 cm
11. Correct to 1 decimal place the volume of a cylinder of height 8 cm and base radius 3 cm is:

[A] 300.0 cm³   [B] 250.0 cm³   [C] 226.2 cm³   [D] 150.9 cm³
12. The following figure shows a rectangular sheet of thin metal from which a cylinder, 10 cm high, is to be made with no overlap. The radius of this cylinder is:

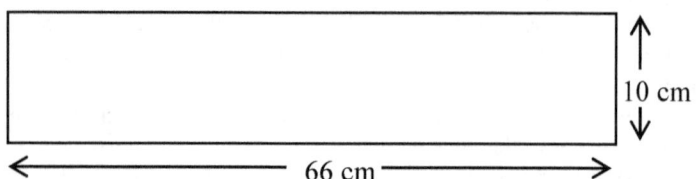

[A] 3.3 cm   [B] 6.6 cm   [C] 10.5 cm   [D] 21 cm
13. A solid cylinder of radius 7 cm is 10 cm long. Its total surface area is:
    [A] 70π cm²   [B] 18π cm²
    [C] 210π cm²   [D] 238π cm²
14. The volume of a cylinder of radius 14 cm is 210 cm³. The curved surface area of the cylinder is:
    [A] 30 cm²   [B] 15 cm²   [C] 616 cm²   [D] 1262 cm²
15. The internal and external radii of a cylindrical bronze pipe are 1.5 cm and 2 cm respectively. If the pipe is 10 cm long, the volume of the bronze used is:
    [A] $5\tfrac{1}{2}$ cm³   [B] 55 cm³   [C] $196\tfrac{2}{5}$ cm³   [D] 550 cm³
16. The cross-section of a prism is a right angled triangle 3 cm by 4 cm by 5 cm. The height of the prism is 8 cm. Its volume is:
    [A] 48 cm³   [B] 60 cm³
    [C] 96 cm³   [D] 120 cm³
17. The following figure shows a triangular prism of length 7cm. The right angled triangle PQR is a cross section of the prism $PR = 5$ cm and $RQ = 3$ cm. The area of the cross-section is:
    [A] 4 cm²   [B] 6 cm²   [C] 15 cm²   [D] 20 cm²

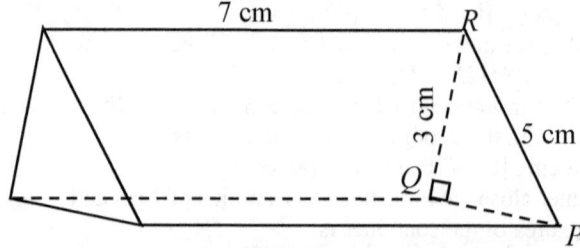

18. The figure above shows a triangular prism of length 7cm. The right angled triangle PQR is a cross section of the prism $PR = 5$ cm and $RQ = 3$ cm. The volume of the prism is:
    [A] 28 cm³   [B] 42 cm³   [C] 70 cm³   [D] 84 cm³
19. A prism is solid figure with:
    [A] regular faces   [B] uniform cross-sectional area.
    [C] triangular faces   [D] a square base and regular triangular faces.
20. The figure that is certainly not a prism is:

Module 7, Topic 9: Prisms and Cylinders

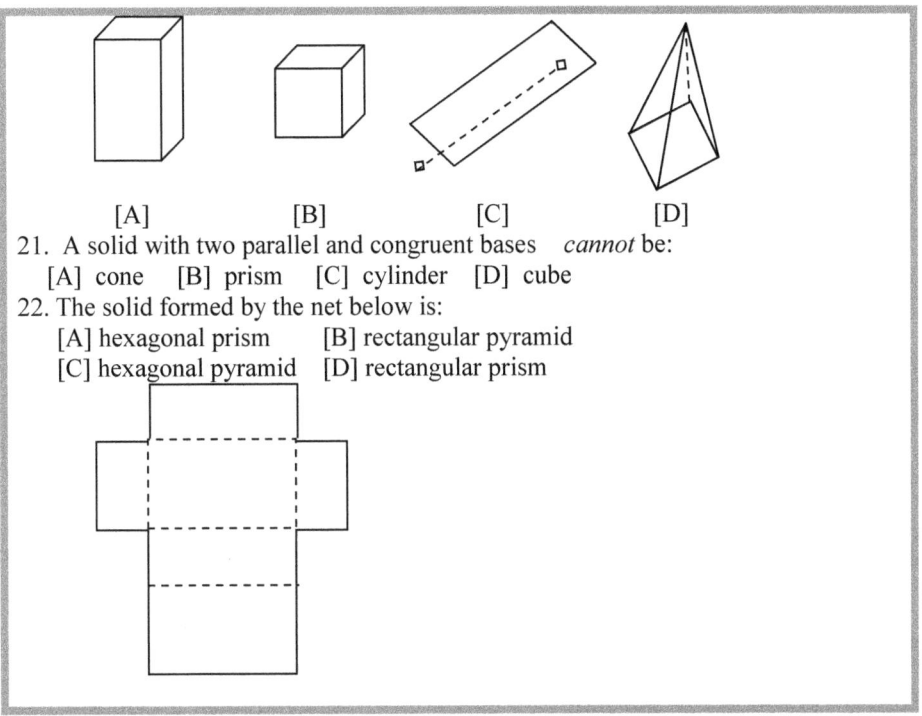

      [A]   [B]   [C]   [D]

21. A solid with two parallel and congruent bases *cannot* be:
 [A] cone [B] prism [C] cylinder [D] cube

22. The solid formed by the net below is:
 [A] hexagonal prism [B] rectangular pyramid
 [C] hexagonal pyramid [D] rectangular prism

Topic 10

PYRAMIDS

Objectives

At the end of this topic, the learner should be able to:

1. Observe and describe a regular pyramid.
2. Recognize and identify a pyramid.
3. Identify the apex, lateral surface, lateral edge, altitude of a pyramid.
4. State the properties of a regular pyramid.
5. Make sketches of a pyramid.
6. Make nets of a pyramid.
7. Make models of pyramid from nets.
8. Use the various parts of the net to establish the original pyramid.
9. Recognize a regular tetrahedron as a special pyramid.
10. Calculate total surface area and volume of a pyramid.

10.1 Observation and Description of Pyramids.

A **pyramid** is a solid with triangular faces and a polygonal base.

  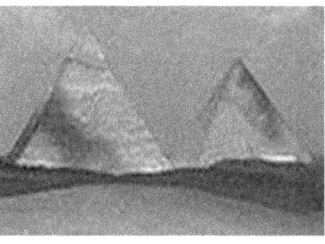

Construction of the pyramids at Giza, Egypt, required knowledge of mathematics. All Egyptian pyramids were aligned to the cardinal points, so that their sides faced due north, south, east, and west. In addition, each side had to slant inward at the same angle and had to narrow toward the top by the same amount so that the four sides met at a single point at the apex.

Karen Petersen

10.2 Naming Pyramids

Pyramids derive their names from the nature of their bases as shown in the table below.

Nature of Base	Name of Pyramid	Diagram
Triangle base	Triangular Pyramid	
Square base	Square pyramid	
Rectangle base	Rectangular pyramid	

Pentagon base	Pentagonal pyramid	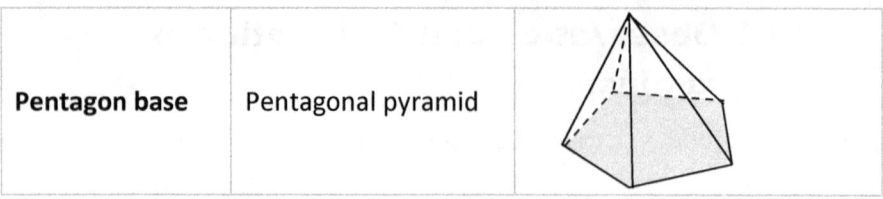

Below is a rectangular pyramid with a base $WXYZ$ and a vertex A, which is not in the same plane with the vertices W, X, Y and Z. The vertex A is called the apex. The pyramid has triangular faces AXW, AXY, AYZ, AWZ. The rectangular base $WXYZ$ is also one of the faces. The edges of the pyramid are $AX, AY, AW, AZ, XY, YZ, XW$ and WZ.

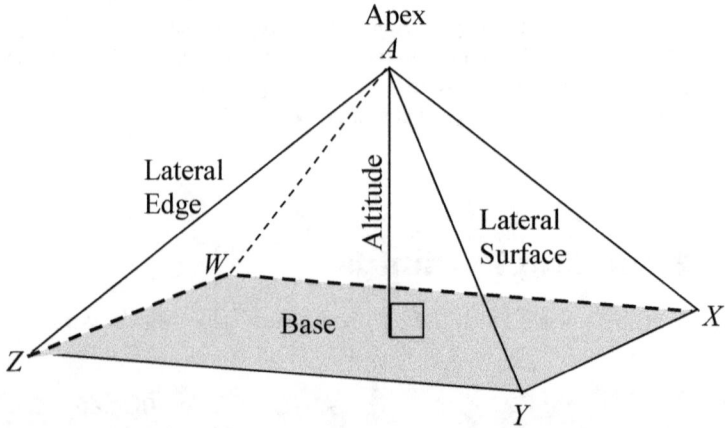

A pyramid is called a **right pyramid** if the line connecting the apex and the center of the base is perpendicular to the base as shown in figure (i) below; otherwise, it is called an oblique pyramid as shown in figure (ii) below.

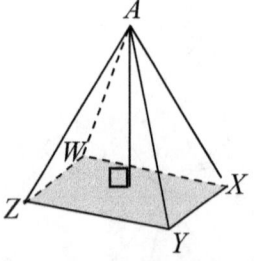 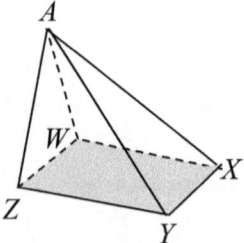

(i) Right rectangular pyramid (ii) Oblique rectangular pyramid

10.3 Polyhedrons

Polyhedrons are solid figures whose faces are made up of polygons. The faces of a regular polyhedron are equal in size and shape. There are only five regular polyhedrons. These are shown below, with their names and number of sides. The commonest of the five regular polyhedrons are the cube (which is also a prism) and the tetrahedron which is a regular triangular pyramid.

A tetrahedron is a solid figure with four triangular congruent faces. A tetrahedron can be looked upon as a special pyramid because it has triangular faces which are congruent (i.e. the faces have the same shape and size).

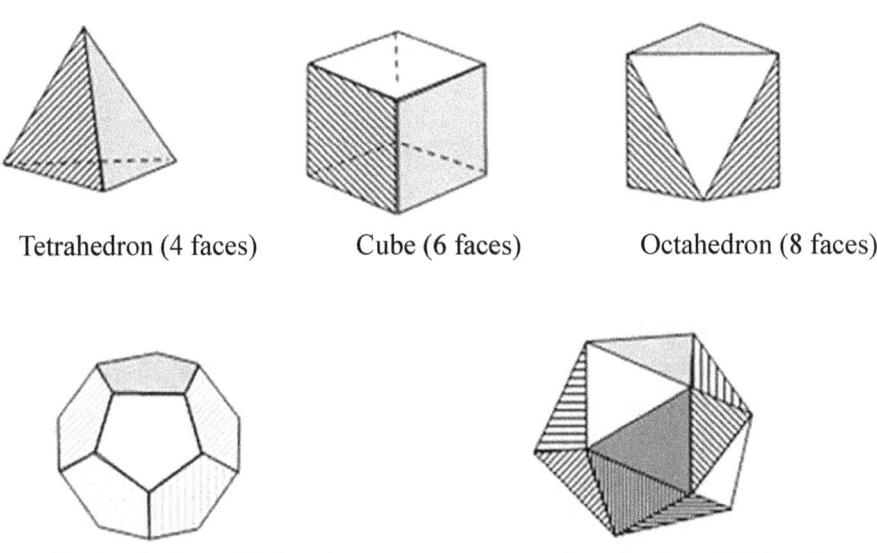

Tetrahedron (4 faces)   Cube (6 faces)   Octahedron (8 faces)

Dodecahedron (12 faces)   Icosahedron (20 faces)

10.4 Mensuration of Pyramids

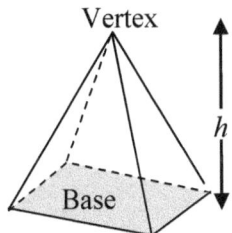

10.5 Volume of Pyramids

The volume V of pyramid with base area A and vertical height h is $\frac{1}{3}$ that of a

prism with the same base area and vertical height. Hence,
$V = \frac{1}{3}Ah$ or $V = \frac{1}{3}$ base area × height

 Example

The base of a pyramid is a square of side 5 cm. Find its volume if its height is 12 cm.

Solution
$V = \frac{1}{3}Ah = \frac{1}{3}(5 \text{ cm})^2(12 \text{ cm}) = 100 \text{ cm}^3$

Surface Area of a Pyramid

To calculate the surface area S of a pyramid, calculate the area of all the faces independently and add to the area of the base. Thus Surface area S of pyramid is given by

S = Area of base + sum of area of all the faces.

 Example

The faces of a pyramid are made of 4 isosceles triangles each with a base 8 cm and height 12 cm. Calculate the surface area of the pyramid.

Solution

The pyramid and one of its faces are as shown below.

Area of each triangle = $\frac{1}{2}bh = \frac{1}{2}(8)(12) = 48 \text{ cm}^2$

Area of faces = $(4)(48) = 192 \text{ cm}^2$

Area of base = $(8 \text{ cm})^2 = 64 \text{ cm}^2$
Surface area = Area of base + Area of faces

$= 64 \text{ cm}^2 + 192 \text{ cm}^2 = 256 \text{ cm}^2$

Module 7, Topic 10: Pyramids

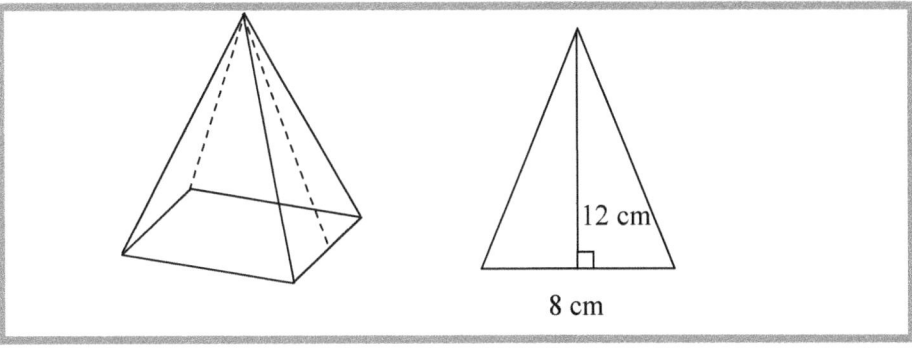

 Exercise 10

1. The base of a pyramid is a square. Each face is an isosceles triangle with base 12 cm and height 16 cm. Calculate
   (a) its surface area   (b) its volume
2. The following figure shows a right rectangular pyramid with base 12 cm by 6 cm. The height of the pyramid is 8 cm. Calculate
   (a) its surface area   (b) its volume

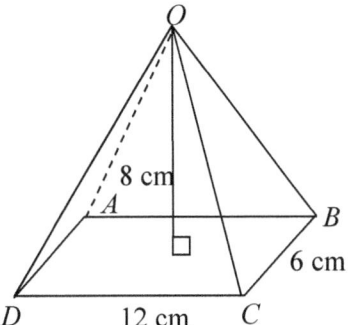

3. The following figure is the net of a model rectangular pyramidal roof. The length and width of the pyramid are 50 cm and 40 cm respectively and the height is 30 cm. Draw and label the heights h_1 and h_2 of the roof. Also determine the surface area of the material which will be used for the model.

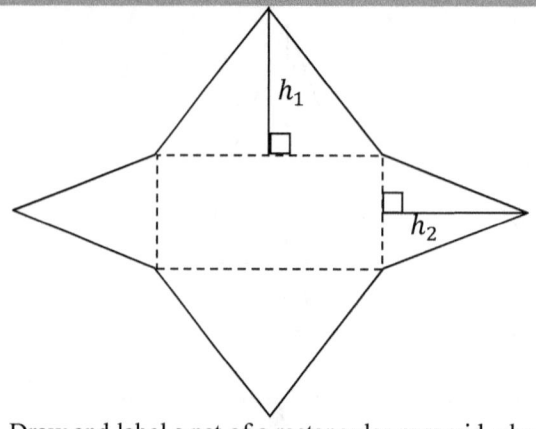

4. Draw and label a net of a rectangular pyramid whose height is 12 cm and whose length and width are 8 cm by 6 cm.
5. You want to make a funnel of a coffee grinding machine which is in the form of an inverted rectangular pyramid open at the rectangular end. The width has to be 30 cm, the length 40 cm and the height 25 cm. Draw and label a model of the net of the funnel.

Multiple Choice Exercise 10

1. The height of a pyramid on a square base is 15 cm. Given that the volume is 80 cm^3, The length of the side of the base in cm is:
   [A] 3.3   [B] 5.3   [C] 4.0   [D] 8.0
2. The height of a pyramid on a square base is 15 cm. If the volume is 80 cm^3, the area of the square base is:
   [A] 16 cm^2 [B] 9.6 cm^2 [C] 8 cm^2 [D] 25 cm^2
3. A right pyramid is on a square base of side 4 cm. The height of the pyramid is 3 cm. The volume of the pyramid is:
   [A] 9   [B] 4   [C] 8   [D] 16
4. A pyramid on a square base of side 10 cm has a height of 15 cm, its volume must be:
   [A] 150 cm^3 [B] 500 cm^3 [C] 1500 cm^3 [D] 5000 cm^3
5. The base of a pyramid is a 12 cm by 12 cm. If its height is 20 cm, the volume of the pyramid in cm^3 is:
   [A] 960   [B] 80   [C] 1440   [D] 1600
6. The net below is a net of:
   [A] a tetrahedron   [B] a pyramid   [C] a cone   [D] a triangular prism

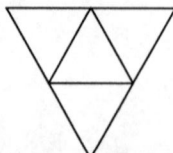

Module 7, Topic 10: Pyramids

7. The following figure is called:

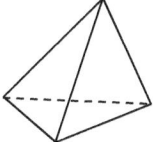

    [A] a triangular pyramid   [B] a triangular prism
    [C] a rhombus   [D] a cone

8. The figure which has one rectangular base and four lateral triangular surfaces is:
    [A] square pyramid   [B] rectangular pyramid
    [C] cone   [D] rectangular prism

9. Neither figure (a) below nor the following diagrams are drawn to scale.

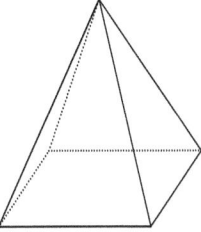

The net which corresponds to the figure above is:

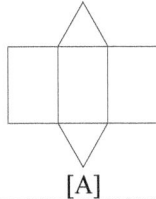   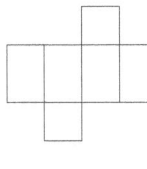   

   [A]   [B]   [C]   [D]

Topic 11
Scales and Similarity

Objectives

At the end of this topic, the learner should be able to:

1. Interpret a given scale to draw on paper.
2. Identify similar figures and corresponding sides.
3. State and use the properties of similar triangles to calculate unknown lengths and angles.
4. Use scales to calculate real/actual lengths.
5. Choose and use appropriate scale.
6. Determine scale factor from corresponding sides.
7. Use scale factor to calculate unknown lengths.
8. Draw to a given scale.
9. Find the sides and /or area of a plane figure from that of a similar figure.
10. Identify congruent figures and their corresponding sides.

Module 7, Topic 11: Scales and Similarity

11.1 Plans and Maps

Before building a house, the plan of the house has to first be drawn. Comparing the size of the paper on which the plan is drawn with the piece of land on which the actual house will be built, it can be seen that the paper is very small. Therefore, the plan has to be small enough to fit on the paper. Also if the plan has to be as big as the house it will not be portable for usage. Maps have to be drawn to fit on paper. Imagine a world map that is drawn to be as large as the surface of the earth. Apart from the fact that it will be very difficult to represent the intended features on such a map, handling will be very difficult and the purpose for which the map was intended will be defeated. Therefore, plans and maps have to be drawn far smaller than the actual area they cover. The relationship between the distance on the plan or map and the actual distance on the earth's surface or object is known as **scale**. In other words a scale is a ratio that compares the length on a drawing to the actual length of an object.

The plan, map or drawing of an object as seen on the paper is called a **scale drawing**.

Scale drawings are very important because many different people use them. These people include builders, surveyors, navigators, engineers, draughtsman, teachers, etc.

Investigative Activity

The figure below shows a triangle ABC and its scale drawing $\triangle A'B'C'$. Measure the angles and lengths of the sides of the two triangles and record your result in tables (a) and (b).

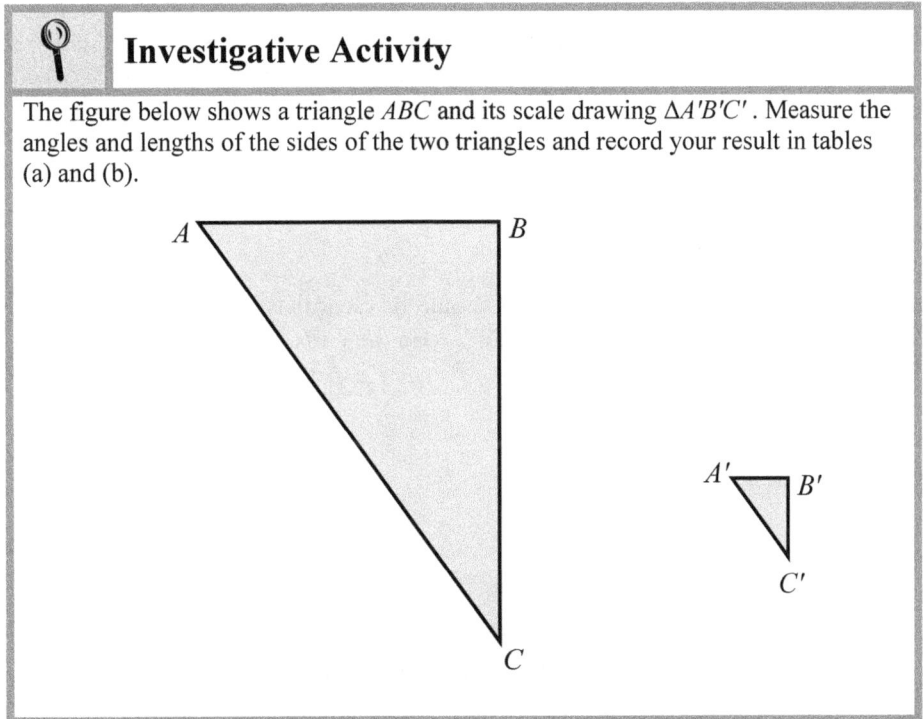

(a)

△ABC		△A'B'C'	
∠ABC		∠A'B'C'	
∠BCA		∠B'C'A'	
∠CAB		∠C'A'B'	

(b)

Side	Length (cm)	Ratio of corresponding sides	
A'B'		$\dfrac{A'B'}{AB}$	
AB			
B'C'		$\dfrac{B'C'}{BC}$	
BC			
A'C'		$\dfrac{A'C'}{AC}$	
AC			

11.2 Similar Figures

In the triangles above, though △ABC and △A'B'C' are different in size, their shape is the same. Two figures which have the same shape but different sizes are said to be **similar** and one is said to be the **enlargement** of the other. Such figures are called **similar figures**. The symbol which is used to denote similar figures is ~. Thus, △ABC ~ △A'B'C'.

Properties of Similar Figures

From our results in the table above, it should be clear that any side on △ABC is 5 times the corresponding side on △A'B'C'. Also the corresponding angles on both figures are equal. Thus, $A\hat{B}C = A'\hat{B}'C'$, $B\hat{C}A = B'\hat{C}'A'$ and $C\hat{A}B = C'\hat{A}'B'$.

From the above we can deduce that,

> Two figures are similar if:
> (i) they have the same shape.
> (ii) corresponding angles are equal.
> (iii) the ratio of their corresponding sides are equal.

Module 7, Topic 11: Scales and Similarity

Scales

It was earlier mentioned that the relationship between the distance on the plan or map and the actual distance on the earth's surface or object is known as **scale**. Also in the triangles above, the ratio of corresponding sides of $\triangle A'B'C'$ to $\triangle ABC$ is $\frac{1}{4}$ since $\frac{1}{4}$ is the scale of the drawing.

Other ways of writing the same scale are 1 is to 4, 1:4 or 1cm represents 4 cm. The number of times that the sides of a figure are multiplied to obtain the sides of its enlargement is called the **scale factor**. Therefore, the scale factor of the enlargement of $\triangle A'B'C'$ to obtain $\triangle ABC$ is 4. A fractional scale factor means that the resulting figure will be smaller in size or reduced. Equally, the larger the scale factor the larger the enlargement.

 Example

1. Given that in the following figure, $\triangle ABC \sim \triangle XYZ$. Find the length of *XY*.

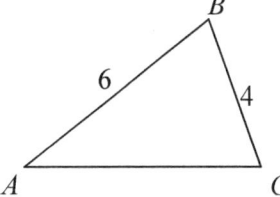 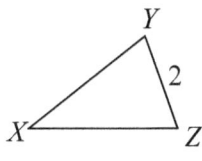

Solution

BC corresponds to *YZ* and *AB* corresponds to *XY*.
BC is divided by 2 to obtain *YZ*.
Therefore, *AB* should be divided by 2 to obtain *XY*.

So, $XY = \frac{6}{2} = 3$ cm.

2. In the following figure, calculate the length of *PQ*.

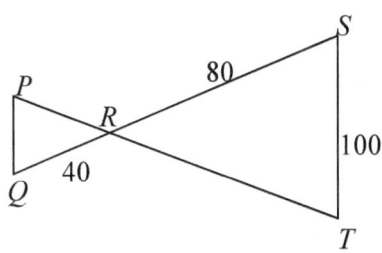

> **Solution**
>
> $\triangle PQR \sim \triangle STR$.
>
> QR corresponds to RS and PQ corresponds to ST.
>
> Also, $QR = \dfrac{RS}{2}$
>
> Therefore, $PQ = \dfrac{ST}{2} = \dfrac{100}{2} = 50$ m.

Choosing an Appropriate Scale

An appropriate scale should:

(i) Be small enough to fit on the drawing paper but big enough to ensure a good degree of accuracy.

(ii) Ensure that calculations involving scale units are as simple as possible. For this reason most good scales are often in powers of 10. For instance,

> 1 centimetre represents 1 unit,
>
> 1 centimetre represents 100 units,
>
> 1 centimetre represents 1000 units, etc.

However, if point (i) will be jeopardized a scale based on factors of powers of 10 should be chosen. For instance,

> 1 centimetre represents 50 units,
>
> 1 centimetre represents 20 units,
>
> 2 centimetres represents 5 units,
>
> 1 centimetre represents 25 units, etc.

Scales involving prime numbers such as 3, 7, 11 etc. which are not powers 10 or factors of powers 10 should as a matter of fact be avoided completely.

Module 7, Topic 11: Scales and Similarity

 Exercise 11:1

1. The scale of a map is 1:200, 000. Find the actual distance which corresponds to the following distances on the map. (a) 6 cm (b) $7\frac{1}{2}$ cm (c) 12.6 cm
2. On a scale drawing of a plane 9 cm corresponds to 108 m on the real plane.
   (a) Find the scale of the drawing.
   (b) Given that the length of the plane is 120 m, find the length of the scale drawing.
   (c) Calculate the actual width if the width on the scale drawing is 7 cm.
3. Njong wants to make a scale model of an elephant 7 m long, 3m high, with teeth of length 15 cm. The model should be no more than 20 cm tall. What scale should he use? Explain.
4. Make a scale drawing of the following rectangle with a scale factor of 1:4.

5. a rectangular farm 110m by 64 m is partitioned into 4 equal portions for the planting of beams, maize, potatoes and vegetables. Using a scale of 1 cm to represent 10 m, draw a scale diagram of the farm showing its partitions.
6. Explain why or why not the pair of triangles is similar.

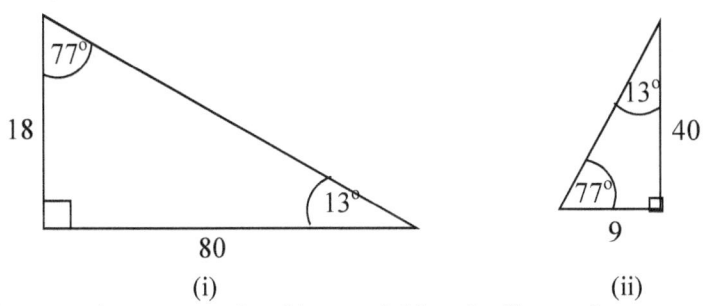

(i) (ii)

7. On a map drawn to a scale of 1 cm to 2.5 km, the distance between two towns is 12 cm.
   (a) Find the actual distance on land.
   (b) The actual distance between two cities is 4 km.
      Calculate the corresponding distance on the map.
   (c) The map is redrawn so that a distance measuring 4 cm now measures 10 cm.

Find the new scale of the map.
8. Explain whether or not all right isosceles triangles are similar.
9. An engineer is drawing plans for a water tower. The tower is 41 m tall and the tank is circular with a diameter of 13 m.
   (a) The engineer builds a model of the tower with a scale of 1 cm: 5 m. What are the dimensions of the model?
   (b) Suppose the engineer decides to build a second model such that the height of model is 10 cm. What is the scale for the model?
10. The following are the designs of two similar trapezoidal gardens. Showing all necessary working, find the perimeter of the smaller garden.

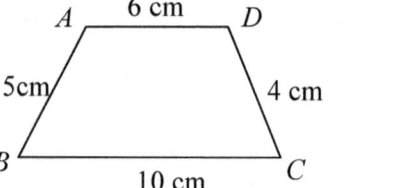 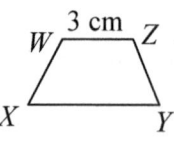

11.3 Congruent Figures

Congruent figures are figures that have the same shape and the same size. Congruent plane figures are figures, which can fit on each other. The statement 'A is congruent to B' is written $A \equiv B$.

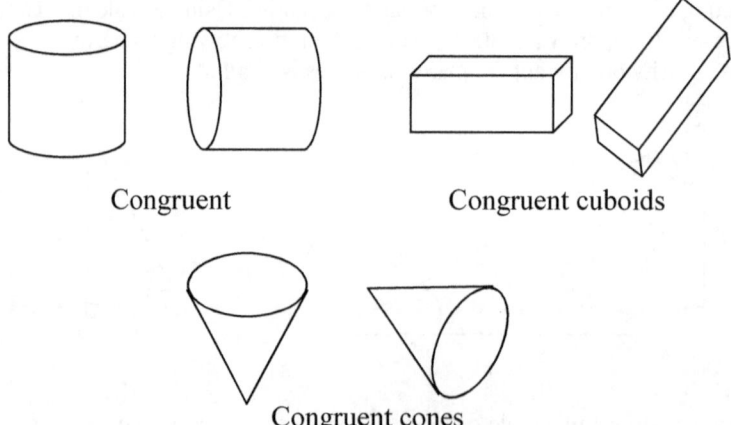

Each pair of solids in the figure above are congruent because they have the same shape and size.

Module 7, Topic 11: Scales and Similarity

Congruent plane figures

Congruent plane figures can be superimposed on each other. i.e. one can fit exactly on the other.

Determine which of the pair of figures above are congruent.

Exercise 11:2

In problems 1 to 7, determine by inspection and measurement, which pair of figures are congruent.

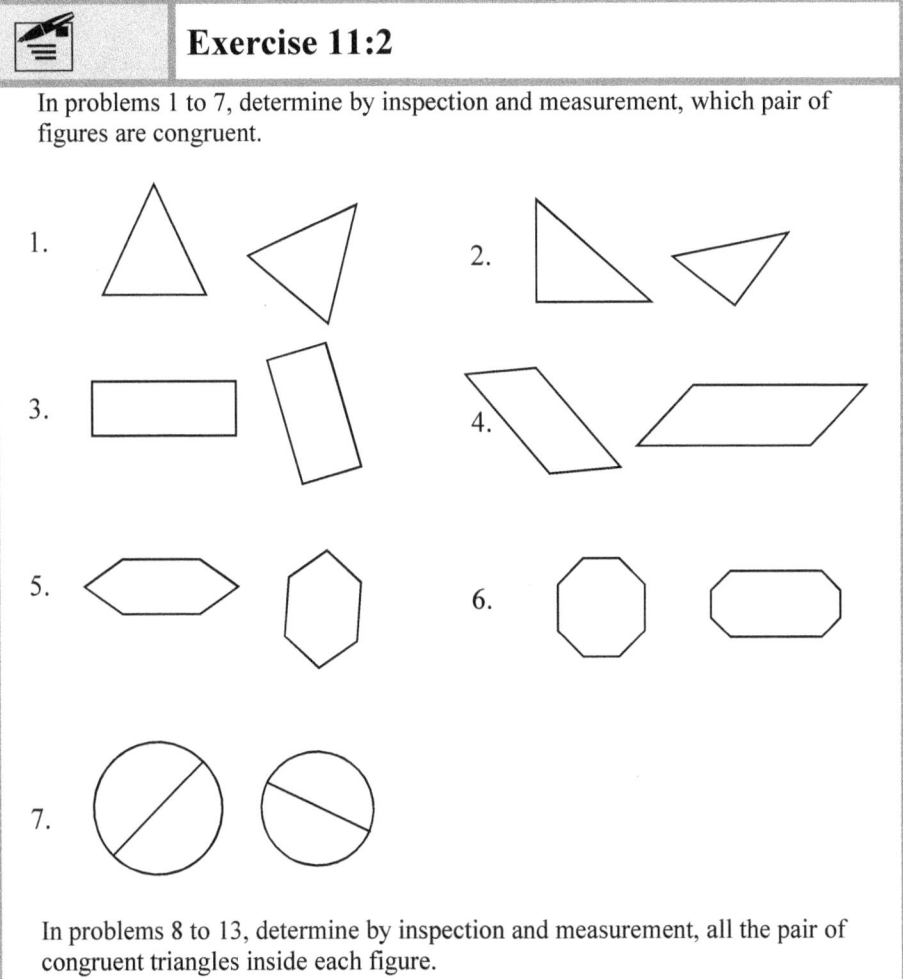

In problems 8 to 13, determine by inspection and measurement, all the pair of congruent triangles inside each figure.

8.

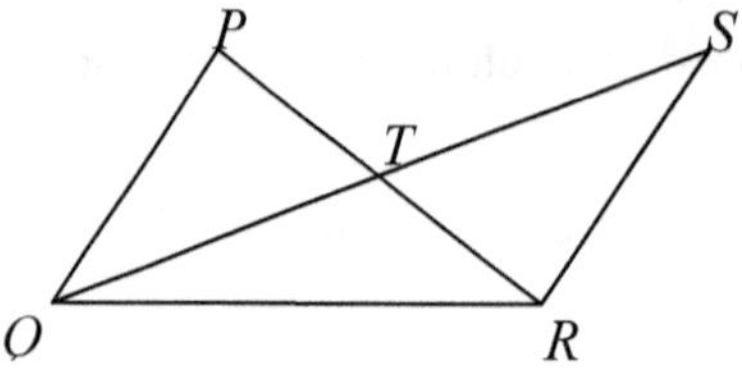

9.

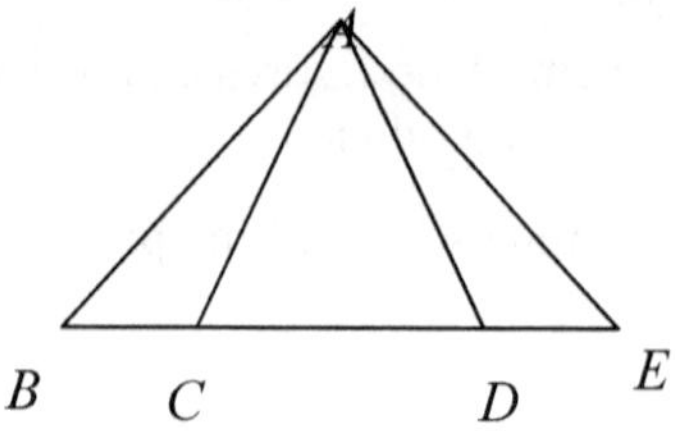

10.

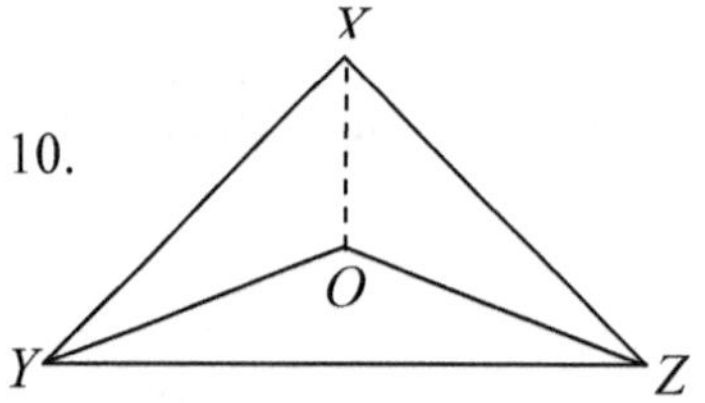

11.

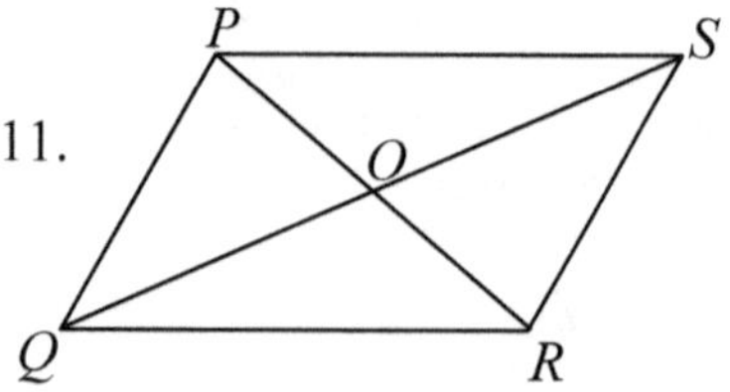

12.

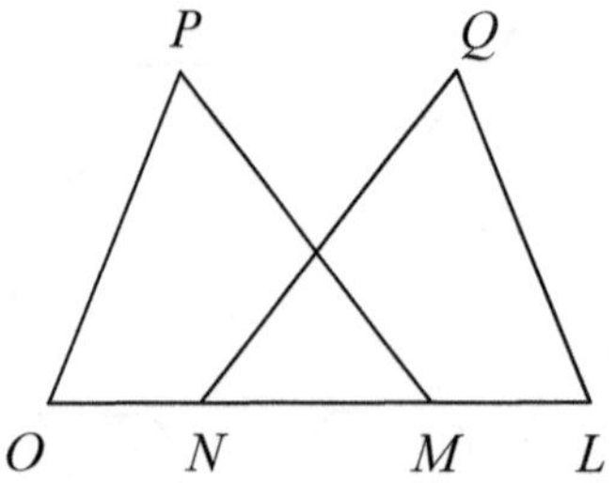

13.

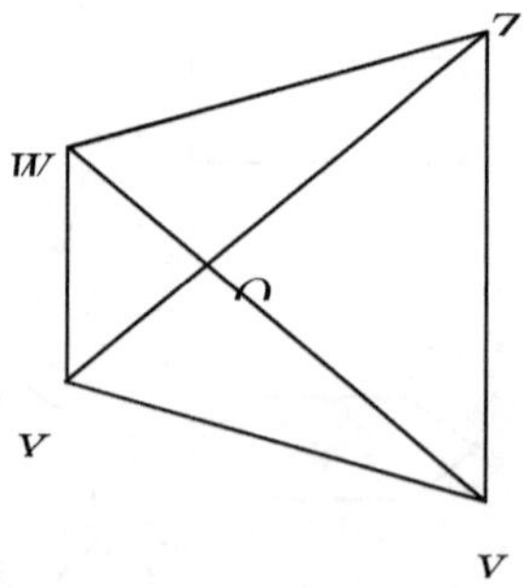

Determine by inspection and measurement whether or not the pair of solids in figure 14 to 16 are congruent.

14.

15.

16.

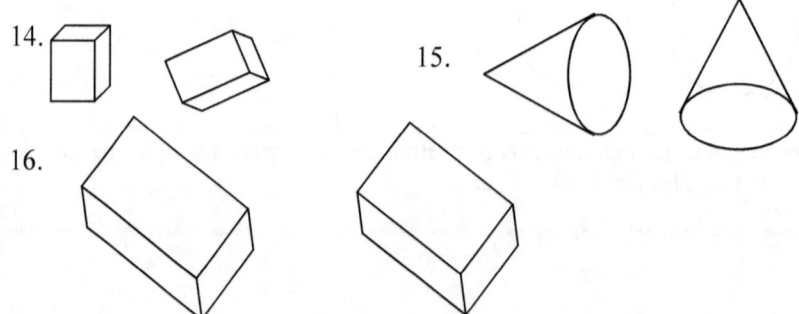

Determine by inspection and measurement whether or not the plane figures in figure 17 to 19 below are congruent.

Module 7, Topic 11: Scales and Similarity

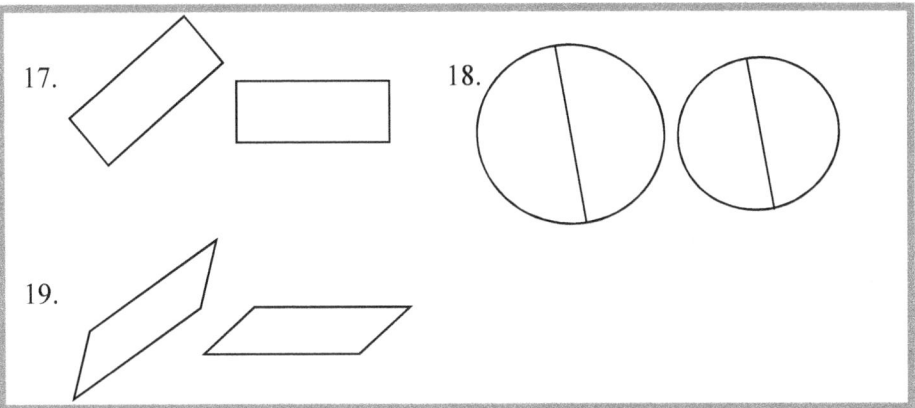

17.

18.

19.

 Integration Activity

The scale drawing above shows a football field.
1. Find the width of the field.
2. Find the length of the shorter side of the penalty area.
3. Find the perimeter of the field.
4. Find the distance from the centre spot to the front of the goal.
5. Song kicks the ball from the penalty kick line to the opposite goal area. Estimate the distance travelled by the ball.
6. Eto'o Fils scores by a direct kick from the point X. Estimate the distance the ball travels before entering the net.

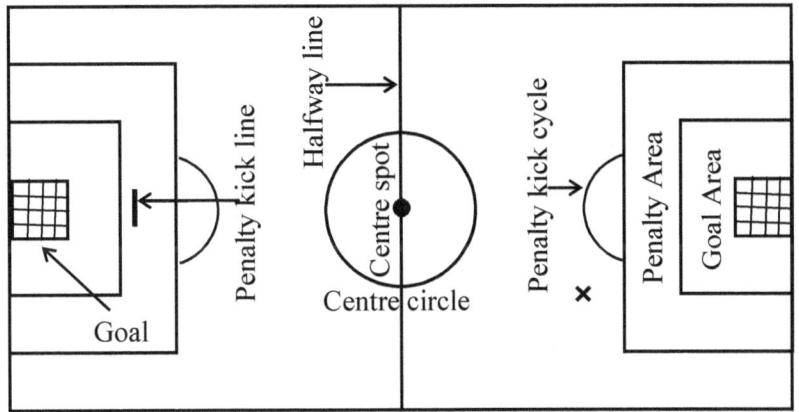

Scale Drawing of a field. 1 cm : 10 m

Competency Base Mathematics for Secondary Schools Book 2

 Multiple Choice Exercise 11

1. The width and length of a plot are in the ratio of 10 to 19. The dimensions of the plot are likely:
   [A] 38 m by 20 m   [B] 20 m by 38 m   [C] 38 m by 2 m   [D] 2m by 38 m
2. A model of a bridge is 24 centimetres long. If the scale of the model is 1 cm : 0.5 m, the length of the bridge is:
   [A] 48 m   [B] 24 m   [C] 6 m   [D] 12 m
3. A map has a scale of 3 centimetres : 8 kilometres. The actual distance between two cities which are 11 centimetres apart on the map, to the nearest tenth of a kilometre is:
   [A] 41.3 km   [B] 29.3 km   [C] 4.1 km   [D] 293.3 km
4. A scale drawing of a market has a scale of 1 cm : 4 m. The actual length of each m in the drawing is:
   [A] 250 m   [B] 25 m   [C] 400 m   [D] 40 m
5. The scale for a model of a plane is 1 mm : 4 cm. The plane is 184 cm long. The length of the plane in the model is:
   [A] 46 cm   [B] 4.6 mm   [C] 181 mm   [D] 46 mm
6. The model of a village is 100 cm long and 70 cm wide. The actual village is 12 km long. The village's actual width to the nearest tenth of a km is:
   [A] 171 km   [B] 58 km   [C] 17.1 km   [D] 8.4 km
7. Using the mm ruler and the map and scale, the approximate distance from Bamenda to Yaoundé if the road were straight is:
   [A] 250 km   [B] 225 km   [C] 290 km   [D] 200 km

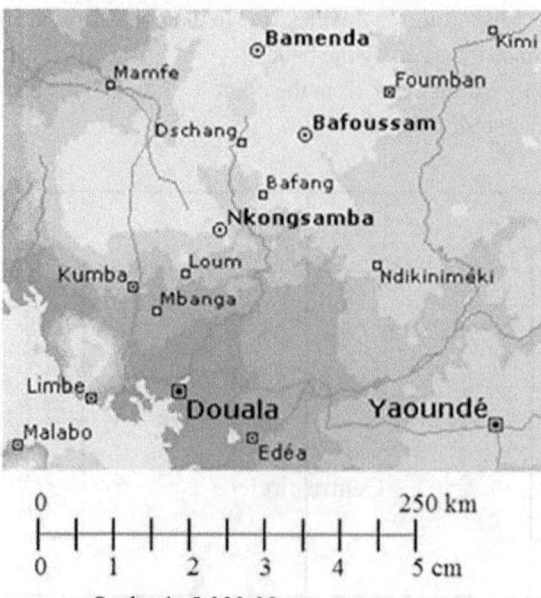

Scale: 1: 5,000,00

8. The pair of polygons is similar. The value of x is:
   [A] 187.5 m   [B] 12 m   [C] 30 m   [D] 37.5 m

Module 7, Topic 11: Scales and Similarity

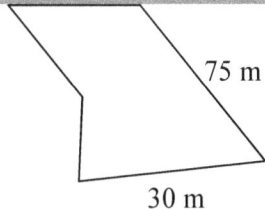 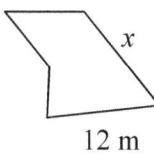

9. △ABC and △DEF are similar. Find the values of x and y.
 [A] x = 7, y = 12 [B] x = 21, y = 12
 [C] x = 14, y = 18 [D] x = 7, y = 18

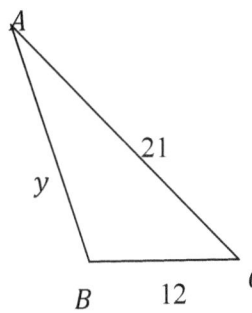 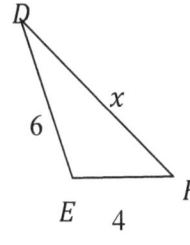

10. The scale on a map is 1 cm : 6 km. Two cities are 13 cm apart on the map. The actual distance between the cities is:
 [A] 13 km   [B] 78 km   [C] 2.17 km   [D] 468 km

11. The width of a picture 20 cm. and that of its reprint is 5 cm. The scale factor for the reprint is:
 [A] 5:1   [B] 1:5   [C] 4:1   [D] 1:4

12. The dotted figure is image of the shaded figure. The scale factor is:

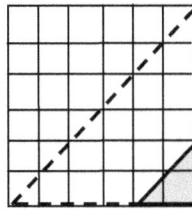

 [A] 2   [B] $\frac{1}{3}$ [C] $\frac{1}{2}$ [D] 3

The diagram below is the plan for a one-bedroom apartment. Use a millimetre ruler for measurements.

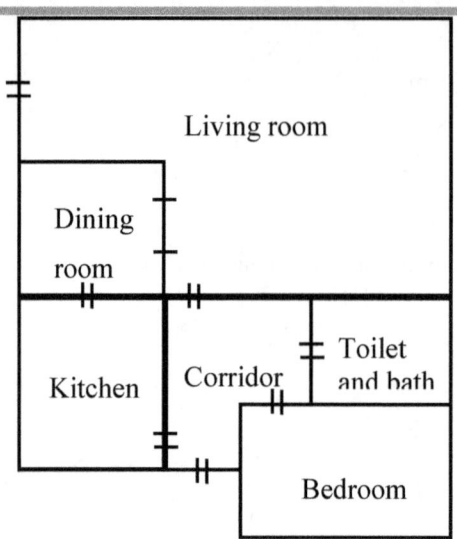

Scale: 1 cm = 1 m

13. The length of the actual kitchen is:
    [A] 2 m   [B] 2.5 m   [C] 3 m   [D] 3.5 m
14. The length of the longest wall in the actual living room is:
    [A] 7 m   [B] 6 m   [C] 5 m   [D] 4m
15. The scale of a map is 1:5,000,000. The number of actual kilometres represented by 22.5 cm is:
    [A] 4.5 km   [B] 45 km
    [C] 112.5 km   [D] 1125 km
16. A woman 160 cm tall casts a shadow 96 cm at the same time her son cast a shadow that 66 cm. The height of the son to the nearest cm is:
    [A] 66 cm   [B] 110 cm   [C] 96 cm   [D] 39.6 cm
17. A building 5 m high casts a 7.5 m shadow at the same time that Nfor casts a 0.6 m shadow. The triangle formed by the building and its shadow is similar to the triangle formed by Nfor and his shadow. The height of Nfor is:
    [A] 0.4 m   [B] 0.5 m   [C] 0.3 m   [D] not here
18. Two ladders are leaning against a wall at the same angle, as shown.

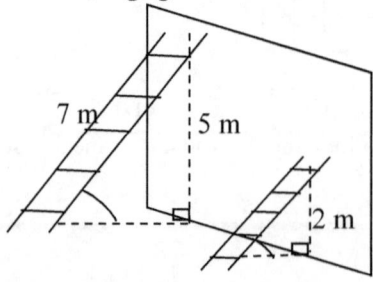

The length of the shorter ladder is:
    [A] 2.8 m   [B] 1.75 m   [C] 2.5 m   [D] 2.0 m
19. To measure the height of a television pole Ngenge placed a mirror on the ground 11 m from the pole, and then walked backward until he was able to see

Module 7, Topic 11: Scales and Similarity

the top of the pole in the mirror. His eyes were 125 cm above the ground and he was 3 m from the mirror. Using similar triangles, the height of the television pole to the nearest hundredth of a metre is:
[A] 55 m [B] 26.4 m [C] 132 m [D] 4.6 m

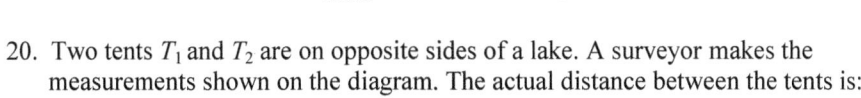

20. Two tents T_1 and T_2 are on opposite sides of a lake. A surveyor makes the measurements shown on the diagram. The actual distance between the tents is:
[A] 314.3 m [B] 648.3 m [C] 642.4 m [D] 311.0 m

21. Given that parallelogram PARL ~ parallelogram WXYZ. The value of c is:

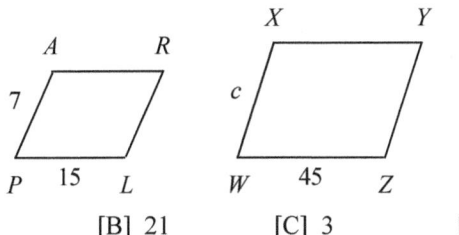

[A] 45 [B] 21 [C] 3 [D] 22

22. To find the distance across the lake in the Figure 20:31, the proportions which can be used is:

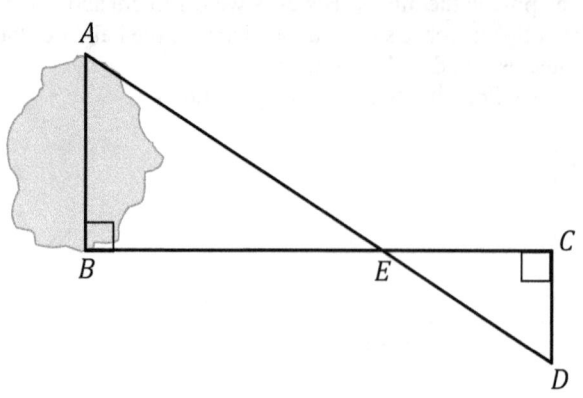

[A] $\frac{BE}{CE} = \frac{AE}{CE}$ [B] $\frac{AB}{CE} = \frac{CD}{BE}$ [C] $\frac{BE}{CE} = \frac{ED}{AB}$ [D] $\frac{BE}{CE} = \frac{AB}{CD}$

23. The triangles below are similar. The value of *a* is:
    [A] 12   [B] 6   [C] 10   [D] 16

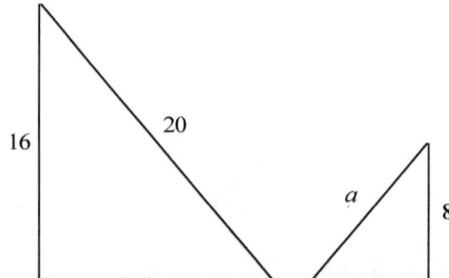

24. The width of the image of a drawing casted by an overhead projector on the screen is 300 mm. The original drawing measures 60 mm wide by 80 mm high. The original drawing and the projected image are similar figures. The height of the projected image is:
    [A] 440 mm   [B] 225 mm   [C] 180 mm   [D] 400 mm

Module 8

Elementary Statistics and Probability

Family of Situations

Module 8 is an extension of module 4 and at the end of the module; the student is expected to acquire many more competencies within the **families of situations** *'Organization of Information and Estimation of Quantities in the Consumption of Goods and Services'*.

Categories of Action

The categories of action for module 8 include:

1. Collection and representation of data,
2. Interpretation of information, predicting and making informed decisions.

Credit

The module is expected to be covered within 3 weeks teaching 4 hours per week (or within 10 to 12 hours).

Topic 12

Representation of Discrete Data

Objectives

At the end of this topic, the learner should be able to:

1. Distinguish between discrete and continuous data.
2. Represent ungrouped data on a frequency distribution table.
3. Represent ungrouped data on statistical graphs such as a bar chart, a pie chart, a line chart and a histogram.
4. Interpret these statistical graphs and make predictions about them.

Module 8, Topic 12: Representation of Discrete Data

12.1 Discrete and Continuous Data

Group Activity

The teacher groups your class into groups of eight to ten students.
(1) Each group is expected to collect the following information concerning members of the class.
   (a) Count the number of students in your group.
   (b) Record the number of prescribed textbooks each student in your group has. You may use the following table to record your results.

Name of student	Number of textbooks

   (c) List the names and ages in years of the students in your group and measure their heights and weights. You may use the following table to record your results.

Name of student	Age	Height	Weight

(2) Assemble the data collected by all the groups of your class in (1) above.

From the above data collection exercise we can see that the data in 1 (a) and (b) can take only exact and separate distinct values. **Discrete data** is data which consists of exact and separate distinct values. Data such as the number of objects is discrete because it can take only the values 0,1,2,3 ... Data such as the suit of playing cards is discrete because it can take only the values clubs, diamonds, hearts and spades.

On the other hand the data in 1 (c) cannot take only exact and separate distinct values. **Continuous data** is data which does not consist of exact and separate or distinct values but consist of a finite or infinite interval. Data such as the heights or weight of objects is continuous because between any two values of such data we can insert other values. For instance we can insert values such as 0.1, 0.2, 0.3 etc. between 1 and 2 in the interval [1,2].

12.2 Frequency Distribution Table for Discrete Data

In section 16.5.2 of book 1 we saw that the number of times an item x occurs is called **frequency** and that a **frequency-distribution table** is a table which shows the frequency f of each item x.

 Exercise 12:1

(1) Draw a frequency-distribution table of the data you collected in the group data in (b) above.
(2) Draw another frequency distribution table which shows the age distribution of the students in your group.

The frequency-distribution table in exercise 12.1 (1) may look something like the following table.

Name of student, x	Number of textbook, f
Giyo	2
Bih	5
Ndoh	3
Abe	4
Che	7
Tsi	9
Nde	1
Ngang	8
Neh	6
Suh	0

Module 8, Topic 12: Representation of Discrete Data

The frequency-distribution table in exercise 12.1 (2) may look something like the following table.

Age, x	Frequency, f
9	5
10	10
11	20
12	15

12.3 Statistical Graphs

In section 16.6 of book 1, we studied some ways of representing data. You are advised to revise that section before continuing.

Example

1. The following table shows the age distribution of students of a certain class. Draw a pie chart to represent this distribution.

Age, x	Nine	Ten	Eleven	Twelve
Percentage	10%	20%	40%	30%

Solution

Age, x	Percentage	Calculation	Angle representing age
Nine	10	$\frac{10}{100} \times 360$	36°
Ten	20	$\frac{20}{100} \times 360$	72°
Eleven	40	$\frac{40}{100} \times 360$	144°
Twelve	30	$\frac{30}{100} \times 360$	108°
TOTAL	100		360°

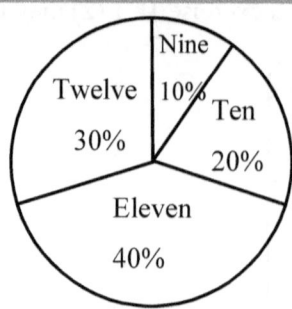

2. In order to find out which type of kola sells best, a businessman decides to keep a record of the number of people who buy four different kolas for one month. The maximum number of people for each kola is 1200 people per month. If less than 25% of the maximum is attained, the businessman will stop the supply of the kola. The following shows a bar chart which the businessman draws at the end of the month. Which of the kola(s) will the businessman stop to supply? Show all working.

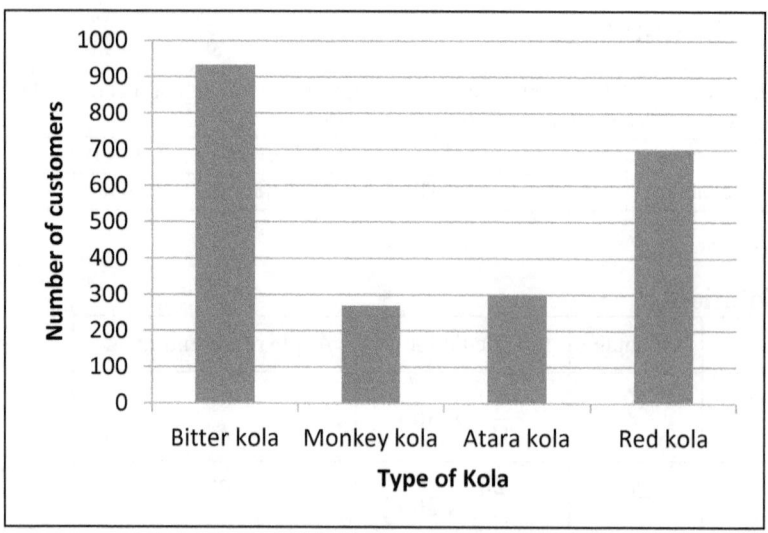

3. The following frequency distribution represents the number of newspapers sold by a newspaper agent on a certain week.

	Number of news papers
Monday	84
Tuesday	46
Wednesday	70
Thursday	82
Friday	98
Saturday	35

Draw a line chart to represent this distribution.

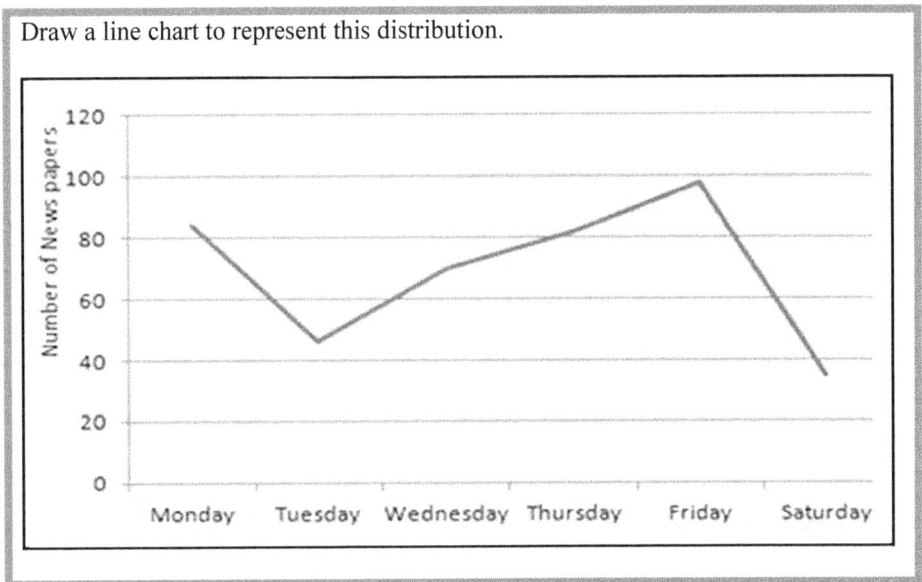

The following table shows the number of students in a certain class who have the required textbooks for the subjects Mathematics (*M*), English (*E*), French (*F*), History (*H*), Geography (*G*), Chemistry (*C*), Physics (*P*), Biology (*B*) and Literature (*L*). Draw a histogram to represent this information.

Textbook	M	E	F	H	G	C	P	B	L
No. of students	14	15	11	3	9	5	7	1	13

Solution

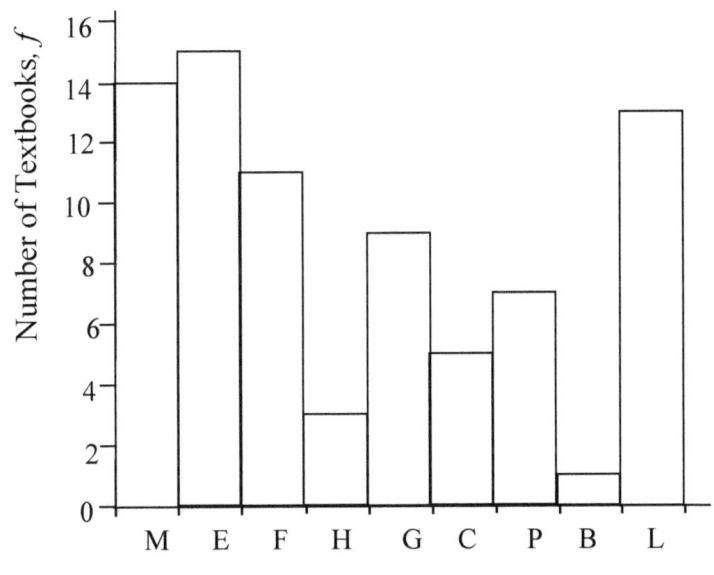

Exercise 12:2

1. The form three students of a certain school participated in team sports as shown in the following table.

Team sport	Number of students
Handball	45
Basketball	60
Football	75

   Represent this information on a pie chart and state the angle for basketball.

2. The figure below is a pie chart (not drawn to scale) showing how a student spent his pocket money amounting to 27,000 FCFA. Given that he spent twice as much on books as he did on taxi, calculate:
   (a) How much he spent on books. (b) How much he spent on others.

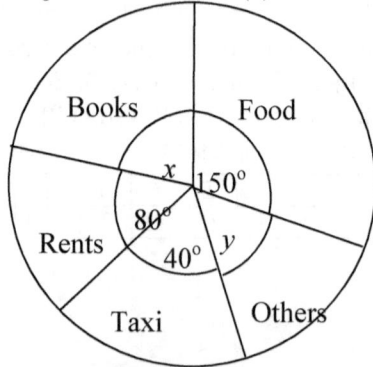

3. The table below shows a survey carried out on a group of students to find out what they ate for launch on a certain day. Draw a histogram to display this data

Achu	15
Rice	9
Garri	4
Bread	2

4. The pie chart below shows the number of votes for candidates A, B and C in an election. Calculate the percentage of the votes to the nearest tenth in favour of candidate B.

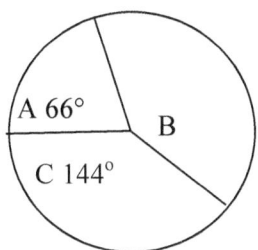

5. Draw a histogram for the following distribution.

Score, x	1	2	3	4	5	6
Frequency, f	4	6	7	3	3	1

6. The livestock of a certain farm consist of 28 cows, 300 sheep, 74 pigs, 306 poultry, 9 dogs and 3 cats. If we are required to record this information on a pie chart, calculate the angle in degrees, at the centre of the sector representing the cows.
7. Five boys A, B, C, D and E are of heights 160, 144, 120, 96 and 80 centimetres respectively. Represent this information on a bar chart.
8. The figure below shows a pie chart indicating the favourite colours of a group of 108 girls.

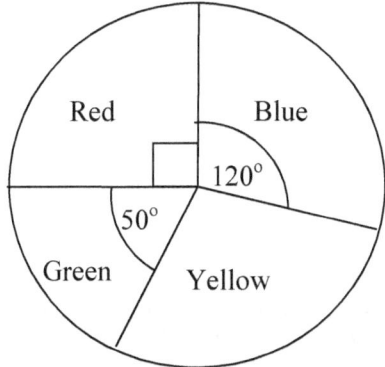

(a) Find the angle of the sector for girls who like yellow.
(b) Find the number of girls who like green.

9. A student used the following table, to draw a pie chart. Find the values of w, x, y and z.

Item type	A	B	C	D
Frequency	48	104	x	y
Sector angle (°)	72	w	108	z

10. The table below shows a statistical table of a variable x with frequency f. We need to represent this data on a pie chart. Calculate in degrees, the angle of the sector representing $x = 3$.

x	1	2	3	4	5	6
f	4	6	7	3	3	1

Multiple Choice Exercise 12

1. A pie chart is drawn to represent the percentages: 20%, 50%, 25% and 5%. The angle which represents 5% is:
   [A] 5°   [B] 18°   [C] 25°   [D] 126°

2. A group of students measured a certain angle (to the nearest degree) and obtained the following results.

   75°   76°   72°   73°   74°   79°   72°
   72°   77°   72°   71°   70°   78°   73°

   The mode of their measurements is:
   [A] 78°   [B] 74°   [C] 73°   [D] 72°

3. The distribution by Region of 840 students in the faculty of science of the University of Buea in a certain session is as follows:

   Adamawa Region   45
   North West Region   410
   Littoral Region   105
   Western Region   126
   South West Region   154

   In a pie chart drawn to represent this distribution, the angle subtended by Western Region is:
   [A] 42°   [B] 45°   [C] 48°   [D] 54°

4. The pie chart below represents the number of fruits on display in a grocery shop. If there are 60 oranges in display, the number of apples are:
   [A] 40   [B] 80   [C] 90   [D] 120

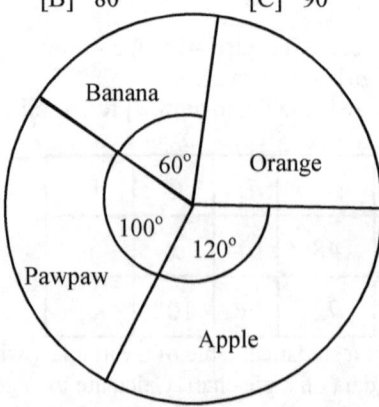

Topic 13

MEASURES OF CENTRAL TENDENCIES

Objectives

At the end of this topic, the learner should be able to:

1. Define the terms mode, median and mean.
2. State the mode of a given discrete data.
3. Determine the median of a given discrete data.
4. Determine the mean of a given discrete data.
5. Interpret information and make predictions related to the information.

13.1 Notion of Measures of Central Tendencies

Mode, mean and **median**, which are very commonly used in statistics, are examples of averages, otherwise called measures of central tendencies. They are so called because their values are representative or typical of any given data, and tend to lie centrally when the data is ranked or arranged in order of magnitude (from highest to smallest or smallest to highest). Each of these measures has its advantages and disadvantages depending on the data and purpose for which it is intended. For instance, the mean has the disadvantage that it is strongly affected by extreme values, while the median is not affected by extreme values. On the other hand, the mode may or may not exist and turns to be very subjective. At times, there are many modes.

13.2 Mode

The mode of any given data is the variable or statistic that occurs most frequently.

 Example

1. Find the mode of the data 1,2,4,6,2,7,7,2,2,7.

    Solution

x	1	2	4	6	7
f	1	4	1	1	3

    2 occur 4 times, which is the highest frequency. Therefore, 2 is the mode.

2. Find the mode of the following data.
    (i) 99, 100, 101, 102 and 101.   (ii) 99, 100, 101, 100, 102 and 101.

    Solution
    (i) Mode =101   (ii) Mode =100 and 101

Notice in Example 2 (ii) above that there are 2 modes, 100 and 101. In such a case, the distribution is said to be **bimodal**. If there are three modes, the distribution is said to be **trimodal** and generally if there are more than one mode the distribution is said to be **multimodal.**

13.3 Median

To obtain the median, first rank the data (or arrange the data in order of magnitude). For an odd number of numbers, the median is the middle number and for an even number of numbers the median is the average of the two middle numbers.

 Example

1. What is the median of 12, 2, 7, 13 and 6?

Solution
Ranking: 2, 6, 7, 12, 13
∴ Median = 7

2. Find the median of 2, 7, 6, 13, 12 and 8.

Solution
Ranking: 2, 6, 7, 8, 12, 13.
Median = $\frac{7+8}{2}$ = 7.5

13.4 Arithmetic Mean (Average or Mean)

The mean is usually denoted by $\bar{x}$, and is obtained by summing all the data and dividing by the frequency. Thus

$$\bar{x} = \frac{\text{Sum of data}}{\text{Total frequency}} \quad \text{or} \quad \bar{x} = \frac{\sum x}{n}$$

Where n is the number of items and $\sum x$ is read 'summation x', meaning 'sum of data' and $\sum$ is called the sigma notation. $\sum x$ can also be read 'sigma x'.

If some of the data repeat themselves, advantage is taken of multiplication as repeated addition, to write the formula as

$$\bar{x} = \frac{\sum fx}{\sum f}$$

$\sum fx$, is read 'sigma fx', or 'summation fx', where fx means the product of each statistic and its frequency.

 Example

1. Find the mean of 11, 9, 15, 12 and 13

Solution

$$\bar{x} = \frac{\sum x}{n} = \frac{11+9+15+12+13}{5} = 12$$

2. The following shows the marks obtained by 30 students during a test. Calculate the average mark.

55	60	65	40	60	60
65	50	40	60	50	60
60	50	60	30	40	60
60	50	60	50	60	50
60	50	60	60	50	60

Solution

To ease the work we draw a frequency distribution table.

Mark, x	Frequency, f	fx
30	1	30
40	3	120
50	8	400
55	1	55
60	15	900
65	2	130
	$\sum f = 30$	$\sum fx = 1635$

$$\bar{x} = \frac{\sum fx}{\sum f} = \frac{1635}{30} = 54.5$$

3. Find the mean of the following data.

13, 13, 13, 13, 13, 13, 14, 14, 15, 15, 15, 16, 16, 16, 16, 16, 16, 16, 16, 16

Solution

x	f	fx
13	6	78
14	2	28
15	3	45
16	9	144
	$\sum f = 20$	$\sum fx = 295$

$$\bar{x} = \frac{\sum fx}{\sum f} = \frac{295}{20} = 14.75$$

Module 8, Topic 13: Measures of Central Tendencies

 Exercise 13:1

1. Find the number that must be removed from the eight numbers 4, 11, 13, 8, 4, 5, 8 and 2, so that the mean of the remaining seven numbers is 6.
2. The mean of five numbers is 4. When a sixth number is added, the mean of the six numbers is $3\frac{1}{2}$. Find the sixth number.
3. Given that the mean of 3, 4 and m is 6, find the mean of 2, m and 14.
4. The following table represents the weights in kg of 11 students.

Weight, kg	45	53	54	49
No. of students	2	3	4	2

   (a) State the modal weight of the students.
   (b) Find the mean weight of the students.
   (c) Find the median of the distribution.
5. Use the frequency distribution in the following table to calculate
   (a) the mean (b) the modal score
   (c) Find the median of the distribution.

Score (x)	1	2	3	4	5	6
Frequency (f)	4	6	7	3	3	1

6. The following table shows the marks obtained by pupils in a mathematics test.
   (a) State the mode of the distribution.
   (b) Calculate, to 1 decimal place, the mean of the distribution.
   (c) Find the median of the distribution.

Marks (x)	0	3	5	6	8	9	10
No. of pupils (f)	2	4	6	2	4	1	1

7. Consider the following frequency distribution.

Score x	3	5	7	9	11
Frequency f	4	6	10	5	5

   (a) State the mode of the distribution.
   (b) Calculate, to 1 decimal place, the mean of the distribution.
   (c) Find the median of the distribution.
8. The following table shows the number of coins of six denominations in a bag. Find
   (a) The average value of the coins in the bag.
   (b) The mode of the coins in the bag.
   (c) The median of the distribution.

Value of coin FRS	5	10	25	50	100	500
Number	4	10	6	8	15	7

9. The weights of 8 teachers in a certain primary school were measured in kg as follows: 74,64,68,76,80,72,68 and 60 respectively. Find
   (a) the median.   (b) the mode of the data.   (c) their mean weight.
10. The frequency distribution below shows the scores in a mathematics test in a certain class.

Score (x)	2	3	4	7	8	9
Frequency (f)	1	4	6	8	9	2

   (a) Find how many students wrote the test.
   (b) Find the mode of this distribution.
   (c) Find the mean mark for the test to 1 decimal place.
11. The numbers of absences from a mathematics class registered within the first 20 lessons in the first term are 2, 3, 1, 0, 0, 4, 3, 2, 2, 2, 1, 4, 5, 5, 0, 0, 1, 1, 2, and 2. Find the
   (a) mode   (b) median   (c) mean number of absences.
12. 10 packets of different sizes contain sweets as shown in Table 22:19.

Number of sweets	5	12	6	15
Number of packets	4	2	3	1

   (a) State the mode of the number of packets.
   (b) Find the median of the number of packets.
   (c) Calculate the mean number of sweets per packet.

Module 8, Topic 13: Measures of Central Tendencies

 Multiple Choice Exercise 13

1. A shoe company will be most interested in the measure of central tendency:
   [A] mean   [B] mode   [C] median   [D] data
2. The average of 0, 1, 6, 7, 9 and 19 is:
   [A] 9   [B] 6   [C] 7   [D] 10
3. The average of 1, 2, 5, 7, and 15 is:
   [A] 6   [B] 30   [C] 7   [D] 15
4. A group of four people measured their heights and found that their heights were 1.38 m, 1.71 m, 1.23 m and 1.40 m. Their average height (in metres) is:
   [A] 1.145   [B] 1.18   [C] 1.39   [D] 1.43
5. The average wage bill in FCFA of 40 men who collectively earn 3,540,000 FCFA is:
   [A] 87,000   [B] 29,500   [C] 88,500   [D] 31,700
6. The mean of 9,13,16,17,19,23,24 correct to two decimal places is:
   [A] 23.00   [B] 17.29   [C] 16.50   [D] 16.33
7. The average of the first four prime numbers greater than 10 is:
   [A] 20   [B] 19   [C] 17   [D] 15
8. The mean of 20 observations in an experiment is 4. If the observed largest value 23 is removed, the mean of the remaining observations is:
   [A] 4   [B] 3   [C] 2.85   [D] 2.60
9. The mean heights of the three groups of students consisting respectively of 20, 16 and 14 students are 1.67 m, 1.50 m and 1.40 m respectively. The mean height of all the students is:
   [A] 1.52 m   [B] 1.53 m   [C] 1.54 m   [D] 1.55 m
10. The mean of 30 observations recorded in an experiment is 5. If the observation largest value of 34 is deleted, the mean of the remaining observations is:
    [A] 4   [B] 3.8   [C] 3.4   [D] 5
11. Given the scores −3, 4,0,4,−2,−5,1,7,10,5 the median of the scores is:
    [A] 2.5   [B] 2   [C] 4   [D] 3.5
12. The median of 8, 10, 9, 6, 7, 10, 12, 8, 9, 8 is:
    [A] 7.5   [B] 8   [C] 8.5   [D] 8.7
13. The median of the set of scores 65, 75, 55, 48, 78 is:
    [A] 55   [B] 60   [C] 72   [D] 65
14. The median of the set of numbers 2.64, 2.50, 2.72, 2.91, 2.35 is:
    [A] 2.72   [B] 2.64   [C] 2.50   [D] 2.35
15. Given the set of numbers 12, 15, 13, 14,12 and 12. The median is:
    [A] 12.5   [B] 12   [C] 13   [D] 13.5
16. The mode of the numbers 8, 10, 9, 9, 10, 8, 11, 8, 10, 9, 8 and 14 is:
    [A] 8   [B] 9   [C] 10   [D] 11

Topic 14

ELEMENTARY PROBABILITY

Objectives

At the end of this topic, the learner should be able to:

1. Define basic probability terms such as event, frequency, likely, certain, uncertain, possible outcomes, impossible outcomes, bias, fair equally likely, chance, sample space, population and probability.
2. Calculate simple probabilities.

14.1 The Concept of Probability

Consider the following statements:

(i) It will probably rain tonight.

(ii) It is likely that the principal will address the students tomorrow.

(iii) The chances of the indomitable lions winning the next world cup football tournament are very high.

How true is each of the above statements? What is the likelihood of it raining tonight? What is the chance of the indomitable lions wining the next world cup football tournament? Questions such as those above lead us into the subject of probability. Probability seeks to answer questions of chance, likelihood, possibility or the degree of truth in an event or something occurring. In other words, probability is a numerical measure of the degree of chance, possibility or likelihood of an event occurring or not occurring.

Probability is often used to take certain life decisions, which depend on chance. For instance, in order to predict the score of say a football march between Egypt and Cameroon, probability comes into play.

14.2 Some Basic Probability Terminology

Suppose a coin is tossed (or thrown), it will either turn up heads (H) or tails (T). If the coin is as likely to turn up heads as to turn up tails then the coin is said to be **fair** or **unbiased**. If the likelihood of the coin turning up heads is not equal to its likelihood of turning up tails the coin is said to be **unfair** or **biased**. The act of tossing the coin is called a **trial** or an **experiment**. The appearance of a head or tail is called an **event**. The event (H or T), one of which must occur in an experiment are called the **outcomes** and the set of all the possible outcomes in a particular experiment is called the **sample space**, usually denoted by S. The sample space S in probability is equivalent to the universal set in set theory.

Thus for the case of the coin, $S = (H, T)$

The set of all possible outcomes in an experiment under specified conditions is called the **event subset** or the **possibility space** and is a subset of the sample space.

 Example

1. An unbiased die is tossed once find the sample space. State the event subset if the event is:
   (i) A: obtaining an even number.

(ii) B: obtaining an odd number.
(iii) C: obtaining a number less than 3.

Solution
$S = \{1,2,3,4,5,6\}$
(a) $A = \{2,4,6\}$
(b) $B = \{1,3,5\}$
(c) $C = \{1,2\}$

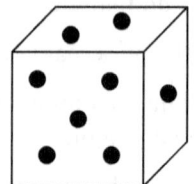

2. A card is drawn at random from a well shuffled pack of 15 cards numbered 1 to 15. Give three event subsets in each case if the event is
   (a) X: obtaining the number 7.
   (b) Y: obtaining a number greater than or equal to 10
   (c) Z: obtaining a multiple of 3

   Solution
   (a) $X = \{7\}$
   (b) $Y = \{10,11,12,13,14,15\}$
   (c) $Z = \{3,6,9,12,15\}$

 ## Exercise 14:1

1. A student has five pairs of socks of the following colours: blue, red, green, white, and black. On one dark morning, the student chooses a pair of socks at random to put on.
   (a) What is the sample space of this experiment?
   (b) What is the event subset for the event that he chooses a white pair of socks or a green pair of socks?
2. Two fair coins are tossed simultaneously. What is the set of possible outcome?
3. A die is tossed. What is the sample space?
4. A bag contains 10 tickets numbered 1-10. State the event subset for the event:
   (a) drawing a prime number,   (b) drawing an even number,
   (c) drawing an odd number,   (d) drawing a square number,
   (e) drawing a multiple of 3.
5. A letter is chosen at random from the letters of the word 'MATHEMATICS'. What is the possibility space for the event of choosing a vowel?
6. A bag contains 4 red marbles, 7 blue marbles and 1 yellow marble. How many elements, has the sample space?

14.3 Probability as a Number

The probabilities that any event E will occur (or will not occur) always lie between zero and one *inclusively*.

This means that the probability of an event occurring is always zero, one or any number between zero and 1. The probability of an event E occurring is denoted by $P(E)$.

Thus; $0 \leq P(E) \leq 1$

The probability of a **sure** or **certain** event A is 1. That is to say

$P(A) = 1 \Leftrightarrow$ Event A must occur.

For instance, the probability that any living person will die one day is 1. The probability of an event B occurring is zero, when it is **impossible** for the event to occur.

$P(B) = 0 \Leftrightarrow$ Event B cannot occur.

For instance the probability that someday a male will be pregnant is zero.

The probability of an event E occurring is any number between 0 and 1 exclusively (not including 0 and 1). This means that there are some chances of the event occurring and some chance of the event not occurring. For instance if it is stated that the probability of an event E occurring is $\frac{1}{4}$ it means that out of 4 trials, the event is expected to occur once and it is expected not to happen 3 times and out of 40 trials, the event is expected to happen 10 times and it is expected not to happen 30 times.

14.4 Equiprobable Outcomes

Equiprobable or **equally likely** events are events which have equal chances of occurrence. If there are n such events, the probability $P(E)$ of one of the events occurring is given by

$$P(E) = \frac{1}{n}$$

> **Example**
>
> 1. A fair coin is tossed once. State the probability of:
>   (a) a head (b) a tail
>
>   **Solution**
>
> $S = \{H, T\}$
>
>   (a) Therefore $P(H) = \frac{1}{2}$ (b) Therefore $P(T) = \frac{1}{2}$
>
> $\Rightarrow P(H) = P(T) = \frac{1}{2}$ or 0.5 or 50%
>
> 2. State the probability of each of the faces showing 1, 2, 3, 4, 5, and 6 if a fair die is tossed.
>
>   **Solution**
> $S = \{1,2,3,4,5,6\}$
> $\Rightarrow P(1) = P(2) = P(3) = P(4) = P(5) = P(6) = \frac{1}{6}$

14.5 Standard Definition of Probability

Suppose a sample space S consists of a finite number of equiprobable outcomes, then the probability of an event E occurring is defined as:

$$\text{Probability of } E = \frac{\text{No of outcomes in the event } E}{\text{Total number of outcomes } S} \quad \text{i.e.} \quad P(E) = \frac{n(E)}{n(S)}$$

Suits of Playing Cards

Ace of hearts   Ace of clubs   Ace of diamonds   Ace of spades

Module 8, Topic 14: Probability

Jack of hearts   Queen of clubs   King of diamonds   Jack of spades

An ordinary pack of playing cards contains 52 cards. There are four types of cards; hearts, clubs, diamonds and spades; each type having 13 members labeled A, 1, 2, 3, 4, 5, 6, 7, 8, 9, 10, Q, K, and J. Each type of card has 3 picture cards labeled **Q, K** and **J**.

Example

1. A card is picked at random from a well shuffled pack of 52 playing cards. What is the probability that it is   (i) An Ace of heart   (ii) A king.

   Solution

   $n(S) = 52$

   (i) $n(\text{Ace of heart}) = 1$

   $\therefore P(\text{Ace of heart}) = \dfrac{n(\text{Ace of heart})}{n(S)}$

   $= \dfrac{1}{52}$

   (ii) $n(\text{king}) = 4$

   $\therefore P(\text{king}) = \dfrac{n(\text{king})}{n(S)} = \dfrac{4}{52} = \dfrac{1}{13}$

2. Fourteen girls are sitting in a circle equally spaced. One is from form four, 2 are from form five, 6 are from lower sixth and 5 are from upper sixth. A girl is selected at random from amongst the girls. Find the probability that the girl is from.
   (i) form four   (ii) form five   (iii) lower sixth   (iv) upper sixth

   Solution
   $n(S) = 14$, $n(\text{form four}) = 1$, $n(\text{form five}) = 2$
   $n(\text{Lower } 6^{th}) = 6$ and $n(\text{upper } 6^{th}) = 5$

   (i) $P(form\ four) = \dfrac{n(form\ four)}{n(S)} = \dfrac{1}{14}$

   (ii) $P(form\ five) = \dfrac{n(form\ five)}{n(S)} = \dfrac{2}{14} = \dfrac{1}{7}$

(iii) $P(\text{lower } 6^{th}) = \dfrac{n(\text{lower } 6^{th})}{n(s)} = \dfrac{6}{14} = \dfrac{3}{7}$

(iv) $P(\text{upper } 6^{th}) = \dfrac{n(\text{upper } 6^{th})}{n(s)} = \dfrac{5}{14}$

 Exercise 14:2

1. A die is tossed. Find the probability of:
    (a) Obtaining a 2.
    (b) Obtaining an even number.
    (c) Obtaining an odd number.
    (d) Obtaining a number less than 5.
    (e) Obtaining a prime number.
2. The following figure shows a spinner. Find the probability of obtaining:
    (a) 3   (b) 1   (c) 2   (d) 5

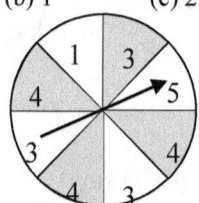

3. In a race of twenty horses, 7 of the horses are black, 8 are white and the rest are dotted. Find the probability that the winner will be dotted.
4. A conference is attended by 9 boys, 12 girls, 15 men and 14 women. Find the probability that a person elected as president will be a man.
5. A letter is chosen at random from the letters of the word 'PROBABILITY'. Find the probability of choosing the letter B.
6. If a number is chosen at random from the integers 5 to 25 inclusive. Find the probability that it is a prime number.

14.6 Complementary Events

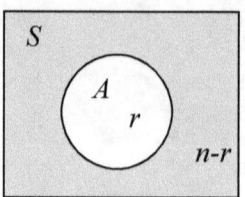

Module 8, Topic 14: Probability

If A is an event in a sample space S, then the event that A does not occur is called "not A" or the "complement of A" denoted by $\bar{A}$ or A' and is defined as the union of all subsets of S whose elements do not belong to A. A' is represented by the shaded portion in the Venn diagram above. Thus if the cardinality of S is n and the cardinality of A is r, then the cardinality of A' is given by

$$n(A') = n(s) - n(A)$$
$$n(A') = n - r$$

Therefore if two events are complementary, the sum of their probabilities is 1.
$$P(A) + P(A') = 1 \Leftrightarrow P(A') = 1 - P(A)$$

N.B.!!

A' is the complement of $A \Leftrightarrow A$ is the complement of A'.

Some examples of complementary events are:

1. The events "obtaining a head" and "obtaining a tail" when a coin is tossed.
2. The events "getting an even number" in one toss of die and "getting an odd number".

 Example

1. The probability that a student will pass the GCE is $\frac{7}{11}$. What is the probability that he will fail?

   Solution
   Since the events 'passing' and 'failing' are complementary,
   $$P(\text{passing}) + P(\text{failimg}) = 1$$
   $$\Rightarrow P(\text{failimg}) = 1 - P(\text{passing}) = 1 - \frac{7}{11} = \frac{4}{11}$$

2. A fair die is tossed find the probability that a 2 will not be obtained.

   Solution
   Let the event of obtaining a two be T, and the event of not obtaining a two be T'.
   $$P(T) = \frac{1}{6}, \quad P(T') = 1 - P(T) = 1 - \frac{1}{6} = \frac{5}{6}$$

Exercise 14:3

1. A bag contains 12 blue marbles, 14 red marbles and 9 green marbles. A marble is drawn at random from the bag. Find the probability that the marble is (a) Green (b) Not blue (c) Red
2. A bag of mangoes contains 30 mangoes 8 of which are bad. A mango is taken at random from the bag. What is the probability that the mango is good?
3. A grocer bought 100 oranges. Given that 16 of them are bad. Find the probability that a mango chosen at random from the lot is good.
4. In a class of 50 students, 35 are girls. Find the probability that a boy will be voted as class prefect.
5. A poultry farm consists of 140 fowls 60 of which are hens. A man buys a fowl from the poultry; find the probability that it is a hen.
6. In a class of 80 students, 75 % of the students pass the examination. Find the probability that a student selected at random from the class failed.

 ### Multiple Choice Exercise 14

1. The probability of having an odd number in a single toss of a fair die is:
   [A] $\frac{2}{3}$ [B] $\frac{1}{6}$ [C] $\frac{1}{3}$ [D] $\frac{1}{2}$
2. The following table gives the scores of a group of students in an English Language test. If a student is chosen at random from the group, the probability that he scored at least 6 marks is:
   [A] $\frac{3}{4}$ [B] $\frac{1}{5}$ [C] $\frac{1}{4}$ [D] $\frac{3}{10}$

Score	2	3	4	5	6	7
Number of students	2	4	7	2	3	2

3. The probability of throwing a number greater than 2 with a single fair die is:
   [A] $\frac{1}{6}$ [B] $\frac{1}{3}$ [C] $\frac{1}{2}$ [D] $\frac{2}{3}$
4. A fair die is rolled once. The probability of obtaining 4 or 6 is:
   [A] $\frac{2}{3}$ [B] $\frac{1}{6}$ [C] $\frac{1}{3}$ [D] $\frac{1}{2}$
5. The probability that an integer selected from the set of integers {20,21,...,30} is a prime number is:
   [A] $\frac{2}{11}$ [B] $\frac{5}{11}$ [C] $\frac{6}{11}$ [D] $\frac{9}{11}$
6. A fair die is rolled once. The probability of obtaining a number less than 3 is:
   [A] $\frac{1}{6}$ [B] $\frac{1}{3}$ [C] $\frac{1}{2}$ [D] $\frac{2}{3}$
7. The data below show the number of workers employed in the various sections of a construction company in Yaoundé. Carpenters 24, Labourers 27, Plumbers 12, Plasterers 15, Painters 9, Messengers 3, and Bricklayers 18. One of the workers is absent on a day. The probability that he is a bricklayer is:

[A] $\frac{1}{9}$ [B] $\frac{2}{9}$ [C] $\frac{1}{6}$ [D] $\frac{1}{4}$

8. From a box containing 2 red, 6 white and 5 blackballs a ball is randomly selected. The probability that the ball selected is black is:

   [A] $\frac{2}{13}$ [B] $\frac{5}{13}$ [C] $\frac{1}{2}$ [D] $\frac{3}{4}$

9. A number is selected at random from the set $Y = \{18,19,20,\ldots,28,29\}$. The probability that the number is a prime number is:

   [A] $\frac{1}{4}$ [B] $\frac{3}{11}$ [C] $\frac{1}{2}$ [D] $\frac{3}{4}$

10. The numbers of goals scored by a school team in 10 netball matches are: 3,5,7,7,8,8,8,11,11,12. The probability that in a match, the school team will score at most 8 goals is:

    [A] $\frac{1}{5}$ [B] $\frac{2}{5}$ [C] $\frac{3}{5}$ [D] $\frac{7}{10}$

Module 9

BASIC ALGEBRA

Family of Situations
At the end of the module 9; the student is expected to acquire competencies within the **families of situations** *'Describing patterns and relationships between quantities using symbols'*.

Categories of Action
The categories of action for module 9 include:
1. Interpretation of algebraic models;
2. Determination of quantities from algebraic models;
3. Representation of quantities and relationships.

Credit
The module is expected to be covered within 2 weeks teaching 4 hours per week (or within 5 to 8 hours).

Topic 15

ALGEBRAIC EXPRESSIONS

Objectives

At the end of this topic, the learner should be able to:

1. Simplify algebraic expressions.
2. Find the value of an expression by substituting numerical values.
3. Bring or combine like terms together.
4. Expand expressions with brackets.
5. Factorize simple expressions.

15.1 Symbolic Expressions

Algebra is a branch of mathematics in which symbols (usually letters) are used to stand for objects or numbers in mathematical expressions and sentences. In algebra we can carry out the basic operations of arithmetic, such as addition, subtraction, multiplication and division, and even many more operations without using specific numbers. Algebra is very useful in everyday life, for calculations and in generalizing situations. Algebraic symbols are used as shorthand symbols to stand for objects. For instance, the statements:

3 books + 2 books = 5 books, can simply be written as $3b + 2b = 5b$.
6 apples − 4 apples = 2 apples, can simply be written as $6p - 4p = 2p$.

Suppose one packet of pens contains 54 pens.
Two packets of pens will contain 2×54 pens.
Three packets of pens will contain 3×54 pens.

Consequently, any number of packets will contain that number multiplied by 54 pens.

We can use any letter to stand for any number.
Hence, n packets of pens will contain $n \times 54$ *pens*.

In algebra, the use of the multiplication sign '×' is generally avoided because it is easily confused with the letter x, which is very often employed. Hence $n \times 54$ is written as $54n$ and is read 'fifty four n'.

In an algebraic product consisting of a numeral and a letter symbol, the numeral is always written before the letter symbol. Therefore, the correct way to write $n \times 54$ in algebra is $54n$ not $n54$. $54n$ means "54 times n".
Thus, if $a = 7$, $2a = a + a = 7+7 = 14$ or $2a = 2 \times 7 = 14$ not 27.
$1b$ is written simply as b, while $-1b$ is written simply as $-b$.

15.2 Variables

A symbol or letter such as n, which stands for any number, is called a **variable**. A variable can be replaced by any number and the statement will still be true. The numbers used to replace or substitute a variable are called the **values** of the variable. Any symbol or letter, which stands for a numeral, is called a **pro-numeral**.

15.3 Algebraic Sentences (Expressions)

Algebraic sentences or expressions are formed by combining pro-numerals and numerals using arithmetical operations such as +, −, x, ÷, square root, and so on.

Module 9, Topic 15: Algebraic Expressions

15.4 From English to Algebra

In translating English sentences to algebraic sentences, certain key words always indicate the operation which is required. The following table is an outline of the most common of the key words.

Addition	Subtraction	Multiplication	Division
Add (to)	Subtract (from)	Multiply (by)	Divide
Sum	Difference	Product (of)	Quotient
Plus	Minus	Times	Share
More than	Less than	Twice, thrice	
Increased by	Decreased by	Of (fractions/percentages)	
	Reduced by		
	Take away		
	Less		

The following table gives some examples of how word expressions can be translated into algebraic expressions.

	Word Expression	Algebraic Expression
1	Multiply a number by 2	$2y$
2	A number increased by 3	$n+3$
3	The sum of 5 and a number	$m+5$
4	4 less than a number	$p-4$
5	Multiply a number by 5 and add 2	$5u+2$
6	3 more than twice a number	$2p+3$
7	A number divided by 4	$\dfrac{x}{4}$
8	The quotient of a number and 7	$\dfrac{x}{7}$
9	The product of 2 numbers	xy
10	A number decreased by 8	$n-8$
11	The quotient of two numbers	$\dfrac{x}{y}$
12	The square of a number	x^2
13	Twice the square of a number	$2x^2$
14	3 less than half a number	$\dfrac{1}{2}x - 3$
15	The difference between two numbers	$b-n$
16	The sum of two numbers	$e+f$

Exercise 15:1

Translate the following into algebraic expressions.
1. Two times a number.
2. Three more than a number.
3. Five less than a number.
4. A number increased by seven.
5. The sum of a number and eight.
6. The difference between a number and 9.
7. Twice a number.
8. Half of a number.
9. A number decreased by ten.
10. Seven more than twice a number.
11. A number divided by two.
12. The product of a number and four.
13. The quotient of a number with 4 as the dividend.
14. The square of a number.
15. The cube of a number.
16. Four less than three times a number.
17. The square root of twice a number.
18. The square of a number increased by two.
19. Fifteen decreased by a number.
20. Two times a number decreased by eight.
21. Seven less than half a number.
22. Half a number increased by seven.

15.5 Variable Substitution

If a and b are variables, then there exist different values of $a + b$ for different values of a and b. The process of finding different values of an expression for different values of the variables is known as **variable substitution**.

 Example

1. Find $a + b$ when $a = 2$ and $b = 8$

   Solution
   When $a = 2$ and $b = 8$, $a + b = 2 + 8 = 10$

2. If $t = s + u$, find t when $s = 6$ and $u = -3$.

   Solution
   $t = s + u = 6 + (-3) = 3$

Exercise 15:2

1. Find the value of $2x + 1$, when $x = 3$.
2. If $y = -1$ and $x = 4$, find the value of $x + 3y$.
3. The area of a rectangle is given by $A = lw$. Find the area when $l = 5$ cm and $w = 2$ cm.
4. What is the difference between $2x$ and 25 when $x = 5$?
5. The perimeter of a field is given by $p = 2(l + w)$. Find p when $l = 8$ and $w = 10$.

In problem 6 and 7 take $\pi = \frac{22}{7}$.

6. The volume, V of a cylinder is given by the formula $V = \pi r^2 h$. Find the volume of a cylinder whose radius r is 2 cm and whose height h is 7 cm.
7. The radius r of a circle is 7 cm. Calculate the area A of the circle if the area is given by πr^2.

15.6 Unknowns and Constants

Consider the statement $3x + 2 = 5$. The only value, which can be substituted for x, for the statement to remain true is 1. Any other value will make the statement false. Such a pronumeral, which can take only particular values, is called an **unknown**. x in this case is not a variable since its value does not change. An unknown may take one or more values. A **constant** is a letter or symbol, which has a fixed value. An unknown can be a constant but a constant is not necessarily an unknown. Examples of constants are; 2, π and -1.

In the formula $A = \pi r^2$, π is a constant but not an unknown.

For a circle with radius $r = 2$ cm, the area A is an unknown since $r = 2$ cm can be substituted in the formula to find A. If on the other hand, there are many different circles with different radii, then both r and A are variables since substituting different values of r will give different values of A.

15.7 Terms and Coefficients

The **terms** of an expression are the different parts of the expression that are linked together by ' + ' or ' − ' signs.

Thus the expression $3xy - 8y + x + 2$ has terms $3xy, -8y, x$ and 2.

An expression with only one term is called a **monomial**, one with two terms a **binomial**, one with three terms a **trinomial** and one with many terms a **polynomial**. A term may simply be a constant called the **constant term** or a variable or a product or quotient of constants and variables.

Each of the numbers or variables that are multiplied together to make up the term is a factor of the term and is called the **coefficient** of the other factors. For instance, the factors of the term $3xy$ are $3, x, y, 3x, 3y, xy$ and $3xy$.

3 is the coefficient of xy and because 3 is a numeral, 3 is called the **numerical coefficient** of xy. On the other hand x, y, xy and are called **literal coefficients** or **non-numerical coefficients** because they are composed of letters.

 Exercise 15:3

(1) Write down the terms of the following expressions:
  (a) $6x - 2xy + 3y$ (b) $px + \frac{p}{y}$ (c) $w + 5pz$ (d) $3pt - \frac{8x}{y} + \frac{1}{w}$

(2) State the numerical coefficients of the following terms:
  (a) $7px$ (b) $-\frac{2}{y}$ (c) $5pw$

(3) What are the literal coefficients of the following terms?
  (a) $3pt$ (b) $-\frac{8x}{y}$ (c) $5pw$

15.8 Like and Unlike Terms

If two or more terms have the same literal coefficient, they are said to be **like terms**. On the other hand, if the literal coefficients of two terms are not the same, the terms are said to be **unlike terms**.

Examples of like terms are:

(a) $3x, 5x$ and $\frac{2x}{7}$ (b) $2pq, \frac{4pq}{3}, -19pq$ (c) $13x, -x - \frac{2x}{3}$

(d) $4ab, \frac{3ab}{5}$ (e) $\frac{ab}{c}, \frac{17ab}{c}, \frac{8ab}{c}$

Examples of unlike terms are:

(a) $4x^2y$ and $4y^2x$ (b) $3x$ and $5y$ (c) $2p^3$ and $2q^3$

15.9 Algebraic Rules

Algebra obeys all the laws of arithmetic and any operation or manipulation, which works in arithmetic, works in algebra. Any operation or manipulation, which does not work in arithmetic, does not work in algebra.

Module 9, Topic 15: Algebraic Expressions

Properties of addition	For all values of x, y, z
Commutative	$x + y = y + x$
Associative	$(x + y) + z = x + (y + z)$
Additive identity property of zero	$x + 0 = x$ and $0 + x = x$

Distributive property	For all values of x, y, z
Over addition	$x(y + z) = xy + xz$
Over subtraction	$x(y - z) = xy - xz$

Properties of multiplication	For all values of x, y, z
Commutative	$xy = yx$
Associative	$(xy)z = x(yz)$
Multiplicative property of zero	$0(x) = 0$ and $x(0) = 0$
Multiplicative identity property of 1	$1(x) = x(1) = x$

15.10   Combining Like Terms

Like terms, of an algebraic expression can be added or subtracted and written as single terms following the algebraic rules. Unlike terms on the other hand cannot be added or subtracted.

Example

1. Compute (a) $a + 2a + 7a$ (b) $5p + \frac{3p}{8} - \frac{5p}{8}$

   Solution
   (a) $a + 2a + 7a = 10a$ (b) $5p + \frac{3p}{8} - \frac{5p}{8} = \frac{38p}{8} = \frac{19p}{4}$

2. Evaluate the following.
   (a) $19x - 15x - 7x + 4x$ (b) $3x + 2y - 5x + 7y$

   Solution
   (a) $19x - 15x - 7x + 4x = x$
   (b) $3x + 2y - 5x + 7y = (3x - 5x) + (2y + 7y) = -2x + 9y$

 Exercise 15:4

Compute the following.
(1) $8x - 4y + 13x - 13y$ (2) $6s + 7t - 8t - 5s$
(3) $12p - 9q - 13q - 7p$ (4) $2x + 7 - 8x - 14 + 5x$
(5) $2u - 5v - u + 2v$ (6) $32a + 6b - 12a + 4b$
(7) $-7x + 8y - 2 + 9x - 10y + 4$ (8) $a + (-a) + b$
(9) $15x - 12y + 88 - 114x + 12y - 8z$
(10) $2pq - 3qr - 7pq + 5rq + 4pq$

15.11 Multiplication and Division of Terms

1. Multiplication

To multiply algebraic terms,
(a) Multiply the numerical coefficients
(b) Multiply the literal coefficients
(c) Write the result as the product of the numerical and literal coefficients placing the numerical coefficient first.

 Example

Evaluate each of the following.

Solution
(i) $(4x)(2x) = 8x^2$ (ii) $(3x)(5y) = 15xy$ (iii) $(-2ab)(3a) = -6a^2b$

2. Division

To divide algebraic terms,
(a) Divide the numerical coefficients
(b) Divide the literal coefficients
(c) Write the result as the product of the numerical and literal coefficients.

Module 9, Topic 15: Algebraic Expressions

 Example

Simplify the following.

(i) $\dfrac{x}{x}$ (ii) $\dfrac{2x^2y}{xy} + \dfrac{5xy}{xy}$ (iii) $\dfrac{20pq}{4p^2}$

Solution

(i) $\dfrac{x}{x} = 1$ (ii) $\dfrac{2x^2y}{xy} + \dfrac{5xy}{xy} = 2x + 5$ (iii) $\dfrac{20pq}{4p^2} = \dfrac{5q}{p}$

 Exercise 15:5

1. Simplify the following.
   (a) $4(2b)$ (b) $3x(x)$ (c) $4p(5q)$ (d) $2ab(9b)$
   (e) $3x(8xy)$ (f) $4xy(7x)$ (g) $10a(5b)$
   (h) $3a(10a)$ (i) $5x(4xy)$ (j) $9uv(3uv)$

2. Simplify the following.
   (a) $8xy \div 4$ (b) $6pq \div q$ (c) $\dfrac{7x^2}{x}$ (d) $\dfrac{54xy}{9y}$
   (e) $\dfrac{33ab}{3a}$ (f) $\dfrac{42pq}{7q}$ (g) $\dfrac{7x^2y}{8x}$ (h) $\dfrac{48u^2v}{12uv}$
   (i) $\dfrac{40xy^2}{8xy}$ (j) $\dfrac{60m^2n}{20mn}$ (k) $\dfrac{12p^2q}{3q^2p} + \dfrac{28pq^2}{7p^2q}$ (l) $\dfrac{15xy}{3y} - \dfrac{24x}{6}$

15.12 Simple Expansions

To expand an algebraic expression, an algebraic product is changed into an algebraic sum.

Expansion of a Monomial and a Binomial

A **binomial expression** is an expression containing exactly two terms (monomials) separated by + or –. $a + b$, $ax + y$, $5x + 7y$, $5y - 2xy$ are examples of binomials.

 Investigative Activity

173

Competency Base Mathematics for Secondary Schools Book 2

> (1) Bame bought 3 pencils at 20 francs each and 3 pens at 50 francs each. Calculate in two different ways, the total cost of the pencils and the pens.
> (2) A man built a store 3 m by 4 m. Later he decided to extend the store by 2 m. Calculate in two different ways the total area of the floor of the store.
> (3) What conclusions do you draw from the solutions in (1) and (2)?

(1) The two methods of solving the problem are outlined below.

Method 1
We may multiply the cost of each by the number of items.
So total cost of the pencils and the pens = $3 \times 20 + 3 \times 50$
$$= 60 + 150 = 210 \text{ Francs.}$$

Method 2
Since the number of pencils is equal to the number of pens, we may add the cost of a pencil and the cost of a pen together and then multiply by 3. So total cost of the pencils and the pens = $3 \times (20 + 50)$
$$= 3 \times 70 = 210 \text{ francs}$$

The following is a sketch showing the original area A_1 and the extension A_2 of the store and the two methods follow.

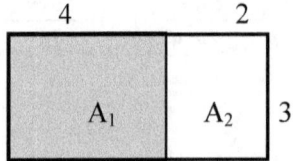

Method 1
We may calculate each area A_1 and A_2 before adding.
So the total area of the floor of the store = $A_1 + A_2$
$$= 4 \times 3 + 2 \times 3 = 12 + 6$$
$$= 18 \text{ m}^2$$

Method 2
We may add the length of the two rectangles (floors) before multiplying by the width.
So the total area of the floor of the store = $(4 + 2) \times 3 = 6 \times 3 = 18 \text{ m}^2$

From the above investigation we can deduced that:
(1) $a(b + c) = ab + ac$ or $(b + c)a = ab + ac$

Module 9, Topic 15: Algebraic Expressions

In like manner,

(2) $a(b-c) = ab - ac$ or $(b-c)a = ab - ac$

These expansions use the **distributive law** and the process is known as **removal of brackets**.

Example

1. Expand the following
   (a) $2(4x + 3)$ (b) $x(2 + y)$ (c) $u(3x - 1)$ (d) $2x(x + 1)$

   Solution
   (a) $2(4x + 3) = 2(4x) + 2(3) = 8x + 6$
   (b) $x(2 + y) = x(2) + xy = 2x + xy$
   (c) $u(3x - 1) = u(3x) - u(1) = 3ux - u$
   (d) $2x(x + 1) = 2x(x) + 2x(1) = 2x^2 + 2x$

2. Expand (a) $(3b - 2c)4$ (b) $(24a - 1)a$

   Solution
   (a) $(3b - 2c)4 = 3b(4) - 2c(4) = 12b - 8c$
   (b) $(24a - 1)a = 24a^2 - a$

Exercise 15:6

Expand the following.
(a) $3(x + 5)$ (b) $(y + 2)4$ (c) $x(4 + y)$ (d) $(3 + u)v$
(e) $2(x - 1)$ (f) $(y - 3)5$ (g) $3p(2 - q)$ (h) $(4 - s)t$
(i) $2x(5 - x)$ (j) $(x - 3)5y$

15.13   Simple Factorization

Factorization is the reverse process of expansion. For instance, since $a(b + c) = ab + ac$, a is the common factor of ab and ac. So to factorise this expression, we remove a from the bracket. Thus,

$$ab + ac = a(b + c)$$

Example

Factories the following: (a) $2x - 2y$ (b) $4xy + 12xz$

Solution
(a) $2x - 2y = 2(x - y)$ (b) $4xy + 12xz = 4x(y + 3z)$

 Exercise 15:7

Factories the following
(1) $3x + 3y$ (2) $2p + 6q$ (3) $5x - 10y$ (4) $4xy + 8y$
(5) $6uv - 3uf$ (6) $\frac{1}{4}xy + \frac{1}{4}px$ (7) $\frac{1}{2}ax - \frac{1}{2}bx$ (8) $\frac{2}{3}y - \frac{1}{3}x$

 Multiple Choice Exercise 15

1. The pair of expressions which are like terms are:
   [A] $2xy$ and $3x$ [B] $5xy^2$ and yx^2
   [C] $3x^2y$ and $7yx^2$ [D] $4y$ and $2yx$
2. The pair of expressions which are unlike terms are:
   [A] $2ab^3$ and $3b^3a$ [B] $5xy^2$ and yx^2
   [C] $3x^2y$ and $7yx^2$ [D] $4xy$ and $2yx$
3. The expressions which are like terms are:
   [A] $\frac{1}{5}xy$ and $\frac{1}{5}x$ [B] $6xy^2$ and $6yx^2$
   [C] $7x^2y$ and $7yx^2$ [D] $6y$ and $6yx$
4. The expressions which are unlike terms are:
   [A] $\frac{1}{7}ab^3$ and $7b^3a$ [B] $7xy^2$ and $7yx^2$
   [C] $7x^2y$ and $7yx^2$ [D] $4xy$ and $2yx$
5. Given that $p = 1, q = -1$ and $r = 0$. The value of $p + q + r$ is:
   [A] -2 [B] -1 [C] 0 [D] 1
6. Given that $p = 1, q = -1$ and $r = 0$. The value of pq is:
   [A] -2 [B] -1 [C] 0 [D] 1
7. Given that $p = 1, q = -1$ and $r = 0$. The value of $p(q + r)$ is:
   [A] -2 [B] -1 [C] 0 [D] 1
8. If $x = 2$, the value of $2x^2 + 3$ is:
   [A] 11 [B] 19 [C] 14 [D] 24
9. When $b = -1$, the value of $5 - b - b^2$ is:
   [A] 5 [B] 3 [C] 14 [D] 24
10. Given $p = 2, q = 5$ and $r = -4$ the value of $3p^2 - q^2 - r^3$ is:
    [A] 51 [B] 19 [C] 2 [D] 0

Module 9, Topic 15: Algebraic Expressions

11. Given that $x = -3$ and $y = -7$, the value of $\frac{x^2-y}{y^2-x}$ is:
    [A] $-\frac{1}{11}$ [B] $\frac{1}{23}$ [C] $\frac{4}{13}$ [D] $\frac{12}{17}$

12. The expression which is not equal to $\frac{1}{2}pq$ is:
    [A] $\frac{pq}{2}$ [B] $\frac{p}{2q}$ [C] $\frac{1}{2}qp$ [D] $\frac{1}{2}p \times q$

13. If $\frac{1}{2}p = x + y$, then p equals:
    [A] $2x + y$ [B] $\frac{1}{2}(x+y)$ [C] $x + 2y$ [D] $2(x+y)$

14. When simplified, $5yx - 7xy + 4yx$ equals:
    [A] $9yx$ [B] $-9xy$ [C] $8xy$ [D] $2xy$

15. The simplified form of $6p + 7q - 8q - 5p$ is:
    [A] $p - q$ [B] $q - p$ [C] $p + q$ [D] $11p - q$

16. $-7x + 8y - 2 + 9x - 10y + 4$ can be simplified to have:
    [A] $2(x - y + 1)$ [B] $2(x - y - 1)$
    [C] $2(x + y - 1)$ [D] $2(x + y + 1)$

17. Simplifying $15x - 12y + 8z - 14x + 12y - 8z$ leads to:
    [A] x [B] y [C] z [D] $x - y + z$

18. By simplification $x + (-x) + y$ is exactly:
    [A] $-y$ [B] y [C] $2x - y$ [D] $2x + y$

19. $32e + 6f - 12e + 4f$ can also be:
    [A] $38e + 6f$ [B] $30e + 8f$ [C] $28e + 9f$ [D] $20e + 10f$

20. When expanded $-2a(3a^2b + 4b^2)$ gives:
    [A] $-6ab^2 - 8a^2b$ [B] $-6ab^2 - 4a^2b$
    [C] $-6ab^2 + 8a^2b$ [D] $-6a^3b - 8ab^2$

21. On Simplification $13x - (2x - 4x - 3x)$ becomes:
    [A] $8x$ [B] $4x$ [C] $-8x$ [D] $-18x$

22. $9x - (5x - 3y) + y$ is equal to:
    [A] $4x - 2y$ [B] $4x + 2y$ [C] $5x - 2y$ [D] $5x + 2y$

23. $(2x + y) + (x - 2y)$ is the same as:
    [A] $3x + y$ [B] $x - 3y$ [C] $x + 3y$ [D] $3x - y$

24. $(2x + y) - (x - 2y)$ is equal to:
    [A] $3x - y$ [B] $x + 3y$ [C] $x - 3y$ [D] $3x + y$

25. $(2x - 3) - (2 - 3x)$ is equal to:
    [A] $5x - 5$ [B] $5x - 1$ [C] $x - 5$ [D] $x - 1$

26. Adding $(2x + y)$ and $(x - 2y)$ gives:
    [A] $3x + y$ [B] $x - 3y$ [C] $x + 3y$ [D] $3x - y$

27. Given the statement $x - 13y + 5z - 4m = x - ($ $)$. The expression required in the bracket is:
    [A] $-13y + 5z - 4m$ [B] $-13y + 5z$
    [C] $-13y + 5z - 4x$ [D] $13y - 5z + 4m$

28. Given that $p = 3 - 2y$ and $q = 4 + 3y$. The value of pq is:
    [A] $-6y^2 - y - 12$ [B] $6y^2 - y - 12$
    [C] $-12 + y + 6y^2$ [D] $12 + y - 6y^2$

Topic 16

Simple Linear Equations and Inequalities

Objectives

At the end of this topic, the learner should be able to:

1. Solve simple linear equations in one unknown.
2. Solve simple linear inequations in one unknown.
3. Solve real life problems that lead to simple Linear equations in one unknown.
4. Represent linear inequations on a straight line.

Module 9, Topic 16: Simple Linear Equations and Inequalities

16.1 The Concept of an equation

An equation is a statement of equality between two expressions. Examples of equations are:

$$2 + 3 = 5 \quad \ldots \ldots \ldots \ldots \ldots \ldots \ldots \ldots \ldots \ldots \quad ①$$

$$2(4.5) = 9 \quad \ldots \ldots \ldots \ldots \ldots \ldots \ldots \ldots \ldots \quad ②$$

$$\frac{n}{6} = 3 \quad \ldots \ldots \ldots \ldots \ldots \ldots \ldots \ldots \ldots \ldots \quad ③$$

$$x + 1 = 4 \quad \ldots \ldots \ldots \ldots \ldots \ldots \ldots \ldots \ldots \ldots \quad ④$$

Though numerical equations such as ① and ② exist, equations usually involve one or more letters or other symbols as in ③ and ④. The letters or symbols are called unknowns and such equations are called **algebraic equations**.

Equations are so important because they are used in almost all branches of pure and applied mathematics and in the physical, biological, and social sciences.

16.2 Simple (Linear) Equations

An equation in which the power(s) of the unknown(s) is (are) unity is called a **linear or simple equation**. Equations ③ and ④ above are examples of linear or simple equations.

To appreciate the use or application of simple equations, consider the following problem.

The cost of a pen is 50 FRS. Nde needed to buy 7 books and 3 pens, but the amount he had was short by 200 FRS, so he decided to buy 5 books and 8 pens for all the money he had. Determine

(a) The price of a book.   (b) The amount he had.

The above problem can very easily be solved if it is translated into a simple algebraic equation. Table 16:1 is a list of some examples of simple equations and their equivalent algebraic translations. Study them carefully and use the skill acquired to translate the simple equation above.

	Word Equation	Algebraic Equation
1	Three more than a certain number is 20.	$x + 3 = 20$
2	Twice a number is 8.	$2n = 8$
3	When a number is multiplied by 5 and 2 is added, the result is 22.	$5y + 2 = 22$
4	A man shares some mangoes equally to his 4 children and each child has 2 mangoes.	$\dfrac{x}{4} = 2$
6	I think of a number, double it and add 1. The result is 7.	$2a + 1 = 7$
7	Njuh buys 2 pens at 50 FRS each and 3 books at y FRS each for 400 FRS.	$100 + 3y = 400$

 Exercise 16:1

Write algebraic equations to represent each of the following:
1. Kanjo is travelling at 3 km/h and Ndi is travelling at x km/h. Their average speed is 4 km/h.
2. A tailor whose daily wage is d FCFA obtains 42,000 FCFA every week.
3. A boy will be 14 years in three years' time.
4. During a football competition a team which played 16 matches altogether, lost three times as many matches as it won.
5. Three times the sum of a number and six is 33.
6. A number increased by 20 equals three times the same number.
7. A number increased by 5 equals twice the same number decreased by 4.
8. Twelve times h equals one hundred and twenty minus thirty-six.
9. MTN charges 60 Frs. per minute for calls. After making a call that last m minutes, Konyuy is charged a bill of 450 Frs.
10. Mbianda bought a radio that costs 16960 FCFA. With tax, t she pays 17810 FCFA.

16.3 Difference between Equations and Expressions

Recall that an **expression** is several terms joined by positive or negative signs. It may consist of only one term. Examples of expressions are $2x + 3y - 4$, $3x + 2$. Expressions do not contain an equal sign (=). An **Equation** on the other hand contains an equal sign. Examples of equations are $3x + 2 = 5$ and $2x + 3y = 4$. In an expression such as $3x + 2$, x is a **variable** but in the equation such as $3x + 2 = 5$, x is an **unknown**. An equation may contain one or more unknowns. For instance, $2x + 3y = 7$ contains two unknowns x and y. Another name for an unknown is the **root of the equation**.

16.4 Additive Inverses

If the sum of two numbers is zero, then one is the **additive inverse** of the other and vice versa. For instance;

The additive inverse of 12 is −12 because $12 + (-12) = 0$.
The additive inverse of −4 is 4 because $(-12) + 4 = 0$.
The additive inverse of $-\frac{1}{2}$ is $\frac{1}{2}$ because $\frac{1}{2} + \left(-\frac{1}{2}\right) = 0$
The additive inverse of $\frac{5}{8}$ is $-\frac{5}{8}$ because $\frac{5}{8} + \left(-\frac{5}{8}\right) = 0$

16.5 Inverse Operations

Inverse operations are operations which can undo each other. Consider the following:

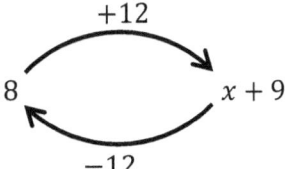

$8 + 12 - 12 = 8$

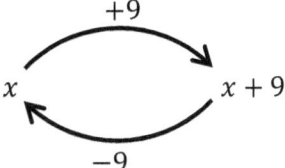

$x + 9 - 9 = x$

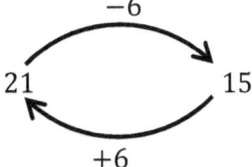

$21 - 6 + 6 = 21$

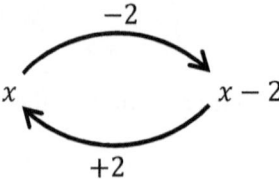
$x - 2 + 2 = x$

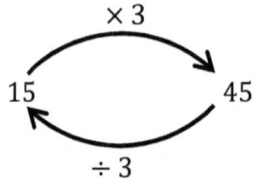
$15 \times 3 \div 3 = 15$

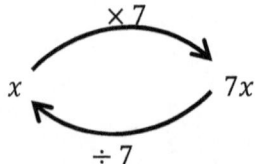
$x \times 7 \div 7 = x$

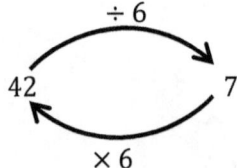
$42 \div 6 \times 6 = 42$

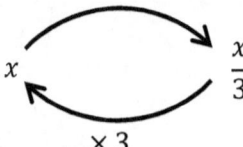
$x \div 3 \times 3 = x$

From the above examples, we can see that:

1. Addition and subtraction undo each other.
2. Multiplication and division undo each other.

Exercise 16:2

1. What number should be subtracted from the expression to have x?
   (a) $x + 5$ (b) $x + 8$
2. What number should be added to the expression to have x?
   (a) $x - 10$ (b) $x - 17$
3. Say what to do so that the result is x in the following cases.
   (a) $3x$ (b) $7x$ (c) $\dfrac{x}{5}$ (d) $\dfrac{x}{4}$
   (e) $0.05x$ (f) $\dfrac{x}{0.02}$ (g) $x + 5$ (h) $\dfrac{x-2}{13}$

16.6 Solving Simple Linear Equations

The process of finding the value of the unknown in an equation or the unknowns in a series of equations is called **solving the equation or equations**. The set of values of the unknown is called the **solution set**. Simple equations can be solved using the **guess method** or the **flow chart method** but the most authentic method for solving equations is the **balancing method**.

The Guess Method

In this method, we speculate the value of the unknown then test it in the equation to see if it is true, if not continue with the speculation.

Example

1. If $2x = 14$, what number gives 14 on multiplying by 2?

   Solution
   Obviously, the number is 7.
       Therefore $x = 7$.
   Checking, $2(7) = 14$ (true)

2. Given that $x - 2 = 9$, what number gives 9 on subtracting 2?

   Solution
   Obviously, the number is 11.
       Therefore $x = 11$.
   Checking, $11 - 2 = 9$ (true)

Competency Base Mathematics for Secondary Schools Book 2

 Exercise 16:3

Solve the following equations using the guess method.
(1) $3x = 18$
(2) $6t = 42$
(3) $\frac{x}{2} = 5$
(4) $\frac{3a}{2} = 12$
(5) $x + 3 = 10$
(6) $y - 4 = 6$
(7) $2 + x = 11$
(8) $16 = 2x + 8$
(9) $2y + 4 = 12$
(10) $7x - 5 = 2x + 11$
(11) $2(n + 5) = 18$
(12) $\frac{2x-5}{3} = 25$

The Balancing Method

When weighing something such as fish or meat on the scale balance, a weight is placed on one side of the balance and an equal weight of fish or meat is then placed on the other side. If a weight of 1kg is placed on one side, 1kg weight worth of fish or meat must be placed on the other side. When the two sides are equal, the scale balance will be horizontal.

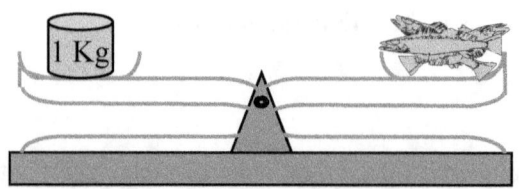

Equations are Balanced Systems

Suppose a 0.5 kg weight is added to the left side, of a scale balance, a 0.5 kg weight worth of fish must be added to the right hand side to maintain equilibrium. Suppose fish of weight 1kg is removed from the right side, the 1kg weight must be removed from the left side to ensure balance. If the weight is tripled, the weight of fish must be tripled to maintain balance.

Thus, an equation is a balanced system and can be compared to a scale balance.

1. Adding the same quantity to both sides does not destroy the balance.
2. Subtracting the same quantity from both sides does not destroy the balance.
3. Multiplying or dividing both sides by the same non-zero number does not destroy the balance.

Module 9, Topic 16: Simple Linear Equations and Inequalities

16.7 One Step Simple Linear Equations

In one step simple linear equations, simply perform one of the following operations to obtain the solution (unknown)

(a) Add the same quantity to both sides to undo subtraction or subtract the same quantity from both sides to undo addition.

(a) Multiply both sides by the same quantity to undo division or divide both sides by the same quantity to undo multiplication.

Example

Solve the following equations:

1. $3x = 18$ 2. $26 = 2p$ 3. $-2y = 6$ 4. $\dfrac{x}{5} = 2$ 5. $\dfrac{t}{6} = -6$

6. $6 = \dfrac{t}{6}$ 7. $\dfrac{2}{3} = \dfrac{x}{15}$ 8. $x + 3 = 9$ 9. $13 = y + 2$ 10. $u - 3 = 11$

Solutions

1. $3x = 18$
    Divide both sides by 3 (or multiply both sides by $\frac{1}{3}$)
    $x = 6$

2. $26 = 2p$
    Divide both sides by 2 (or multiply both sides by $\frac{1}{2}$)
    $p = 13$

3. $-2y = 6$
    Divide both sides by -2 (or multiply both sides by $-\frac{1}{2}$)
    $y = -3$

4. $\dfrac{x}{5} = 2$
    Multiply both sides by 5
    $x = 10$

5. $\dfrac{t}{6} = -6$

Multiply both sides by 6
$t = -36$

6. $6 = \dfrac{t}{6}$
   Multiply both sides by 6
   $t = 36$

7. $\dfrac{2}{3} = \dfrac{x}{15}$
   Multiply both sides by the GCD (i.e. 15)
   $x = 10$

8. $x + 3 = 9$
   Subtract 3 from both sides
   $x = 6$

9. $13 = y + 2$
   Subtract 2 from both sides
   $11 = y$

10. $u - 3 = 11$
    Add 3 to both sides
    $u = 14$

 Exercise 16:4

Solve the following equations.

1. $4x = 32$
2. $15 = 5t$
3. $-3p = 18$
4. $\dfrac{u}{7} = 6$
5. $4 = \dfrac{z}{9}$
6. $\dfrac{w}{-3} = 11$
7. $-9 = \dfrac{r}{3}$
8. $\dfrac{-5q}{2} = 15$

We can also write the solution using set builder notation as follows.

Solution = $\{x: x = 6\}$

Solution = $\{y : y = -3\}$

Solution = $\{q : q = 18\}$

Module 9, Topic 16: Simple Linear Equations and Inequalities

16.8 Two-Step Simple Linear Equations

Most of these types of equation have like terms on both sides. As before, perform the inverse of the operations identified aiming at isolating the unknown.

 Example

Find the solution set of each of the following equations.
(a) $1 - 2x = -7$ (b) $2r - 5 = 8$ (c) $9 = 4m - 7$
(d) $7p = 8 + 3p$ (e) $5x + 9 = 2x$ (f) $\dfrac{v}{4} - 6 = 2$

Solutions
(a) $1 - 2x = -7$
    Subtract 1 from both sides
    $-2x = -8$
    Divide both sides by -2
    $x = 4$
    Solution set $= \{x: x = 4\}$

Explaining the steps taken is not always necessary. Wherever the explanations are given, this is to make the learner understand the procedure. Thus;

(b) $2r - 5 = 8$
    $2r = 13 \Rightarrow r = \dfrac{13}{2}$
    Solution set $= \left\{r : r = \dfrac{13}{2}\right\}$

(c) $9 = 4m - 7$
    $16 = 4m \Rightarrow 4 = m$
    Solution set $= \{m: m = 4\}$

(d) $7p = 8 + 3p$
    $4p = 8 \Rightarrow p = 2$
    Solution set $= \{p: p = 2\}$

(e) $5x + 9 = 2x$
    $9 = -3x \Rightarrow -3 = x$
    Solution set $= \{x: x = -3\}$

(f) $\dfrac{v}{4} - 6 = 2$

$$\frac{v}{4} = 8 \Rightarrow v = 32$$

Solution set = $\{v: v = 32\}$

 Exercise 16:5

Solve the following equations.
1. $2x + 5 = 9$
2. $4t + 11 = 21$
3. $20 = 3q + 8$
4. $13 = 6 + 7p$
5. $2a - 5 = 9$
6. $4x - 11 = 21$
7. $60 = 10d - 20$
8. $11 = 6y - 16$
9. $\frac{u}{4} + 3 = 7$
10. $\frac{m}{5} + 2 = 10$
11. $17 = \frac{n}{2} + 15$
12. $25 = \frac{x}{10} + 2$
13. $\frac{k}{4} - 3 = 7$
14. $\frac{z}{5} - 2 = 10$
15. $2(n + 5) = 18$
16. $5u = 2u + 27$
17. $5x = 40 - 3x$
18. $2x = 90 - 7x$
19. $10t - 11 = 8t$
20. $18 - 5a = a$
21. $9u = 16u - 105$
22. $13y = 15 + 3y$
23. $4y + 5 = 5y - 30$
24. $3t + 10 = 2t + 20$

16.9 Multi-Step Simple Linear Equations

Most of these types of equations have the unknown on both sides. To solve them, perform the inverse of the operations identified aiming at combining like terms and bringing the unknown on one side.

 Example

Solve the equations
(1) $5x - 2 = 3x + 4$
(2) $-7 - 4a = 48 + 7a$

Solutions

(1) $5x - 2 = 3x + 4$
Add 2 to both sides
$5x = 3x + 6$
$2x = 6 \Rightarrow x = 3$

(2) $-7 - 4a = 48 + 7a$
Add 7 to both sides
$-4a = 55 + 7a$
Subtract $7a$ from both sides
$-11a = 55 \Rightarrow a = -5$

Module 9, Topic 16: Simple Linear Equations and Inequalities

Now we have built enough skills to solve Nde's problem.

We were required to find the price of a book and determine the amount Nde had.

$7b + 150 - 200 = 5b + 400$

$\Rightarrow 7b - 50 = 5b + 400$

Adding 50 to both sides,

$7b = 5b + 450$

Subtracting $5b$ from both sides,

$2b = 450$

Dividing both sides by 2,

$b = 225$

The price of a book 225 FRS

The amount Nde had $5b + 400 = 5(225) + 400 = 1525)$ FRS

 Exercise 16:6

Solve the following equations.
1. $5p + 3 - 2p = p + 8$
2. $7m + 10 - 2m = 16m - 12$
3. $9y - 19 = 5y + 21$
4. $3x - 7 = 5 - x$
5. $6x - 3 = 7 + x$
6. $5 - 6x = x - 9$
7. $36 = -2(m + 3) + m$
8. $6 = 2(x + 8) - 5x$
9. $3(t + 7) = t - 19$
10. $5(4x - 2) = x + 9$
11. $-6p - 21 = 3p - 12$
12. $3x + 5 = 21 - x$

16.10 Simple Linear Equations Involving Fractions and Decimals

When an equation involves fractions, multiply both sides by the LCM of the denominators to get rid of the fractions before solving. If it involves decimals, multiply both sides by the power of ten, which has the same number of zeros as the decimal with the highest number of decimal places.

Example

Solve the following equations.
(a) $1.1x + \frac{1}{5}x = 12.2 - 3\frac{1}{10}$ (b) $0.17x - 10.966 = 1\frac{7}{20}x + 36.234$

Solution

(a) $1.1x + \frac{1}{5}x = 12.2 - 3\frac{1}{10}$

$1.1x + \frac{1}{5}x = 12.2 - \frac{31}{10}$

Multiply both sides by 10.

$1.1x + \frac{1}{5}x = 12.2 - \frac{31}{10}$

$11x + 2x = 122 - 31$

$13x = 91$

$\Rightarrow x = 7$

(b) $0.17x - 10.966 = 1\frac{7}{20}x + 36.234$

$0.17x - 10.966 = 1.35x + 36.234$

Multiply both sides by 1,000.

$170x - 10966 = 1350x + 36234$

$-1180x = 47200$

$x = -40$

 Exercise 16:7

Solve the following equations.
1. $0.6x - 0.4 = 0.4x + 0.6$
2. $p - 0.1p + 0.9 = 0.2(p + 1)$
3. $\frac{3}{4}x + \frac{1}{5}x = \frac{1}{2}x - \frac{3}{10}$
4. $\frac{14}{3} - 7x = 8$
5. $3 = \frac{w}{12} - 7\frac{1}{4}$
6. $5\frac{1}{2} = \frac{x}{8} - 4$
7. $\frac{2x-5}{3} = 25$
8. $\frac{4x}{3} + 2 = \frac{5x}{2} - \frac{3}{2}$
9. $3y + \frac{1}{2}y - \frac{2}{5}y = \frac{y}{10} + \frac{7}{10}$
10. $100 + 3\frac{1}{2}x = 23\frac{1}{2}x$

16.11 Word Problems on Simple Linear Equations

Nde's problem is an example of a real life application of simple linear equations. The following are more examples.

> **Real life Examples**
>
> 1. Think of a number, divide it by 15 then subtract 3, the result is $\frac{1}{3}$. What is the number?
>
>   **Solution**
> $$\frac{x}{15} - 3 = \frac{1}{3}$$
>   Multiply both sides by 15.
> $$x - 45 = 5$$
>   Add 45 to both sides
> $$x = 50$$
>
> 2. A girl has three times an amount of money as her friend. If their total sum is 700 FRS, find how much each of them has.
>
>   **Solution**
>   Let the girl's friend have x FRS.
>   Then, the girl has $3x$ FRS.
> $$3x + x = 700$$
> $$4x = 700$$
>   Divide both sides by 4
> $$x = 175$$
>   Therefore, the girl has 525 FRS and her friend has 175 FRS.

 Integration Activity

You have 10000 francs and needs to buy a pair of shoes costing 4100 francs and as many ledgers as possible. You equally need to pay taxi to and from the market. Given that a taxi drop is 250 francs and a ledger cost 700 francs. How many ledgers can you buy so that you should equally have some little amount to eat at school?

Exercise 16:8

1. The length of a rectangle is 2 cm longer than its width. Find the length of the rectangle if the perimeter is 48 cm.
2. A woman bought some mangoes and shared to her four children, each child having two. How many mangoes did she buy?
3. A student buys two pens at 50 FRS each and three exercise books at y FRS each. If her total expenditure on these is 400 FRS, find the value of y.
4. Kanjo and Ndi are travelling at x km/h and $3x$ km/h respectively. Find the value of x if the arithmetic mean of their speeds is 4 km/h.
5. The result of a number divided by two is the same as the difference between 3 and the number divided by 4. What is the number?
6. Ayuk is travelling at 3 km/h and Tabe is travelling at x km/h. Their average speed is 4 km/h. Find the value of x.
7. A tailor whose daily wage is d FCFA obtains 42,000 FCFA every week. What is his daily wage?
8. A boy will be 14 years three years from now. How many years was he 4 years ago?
9. During a football competition a team, which played 16 matches, altogether lost three times as many matches as it won. How many marches did the team win?
10. Three times the sum of a number and six is 33. Find the number.
11. A number increased by 20 equals three times the same number. What is the number?
12. Twelve times h equals one hundred and twenty minus twelve. Find h.
13. MTN charges 60 Frs. per minute for calls. After making an MTN call that last m minutes, MTN charges Konyuy a bill of 420 Frs. For how many minutes did Konyuy make the call?
14. Mbianda bought a radio that costs 16960 FCFA. With tax, t she pays 17810 FCFA. Find the value of t.
15. Tanto intends to buy a bicycle, which costs 27000 FCFA. He has 6000 FCFA and saves 3000 FCFA every week for n weeks. Write down an equation and use it to calculate the number of weeks he saves 3000 FCFA.
16. A number increased by 5 equals twice the same number decreased by 4. Find the number.
17. Use the Pythagoras theorem (stated below), to find the value of x in the given right-angled triangle.

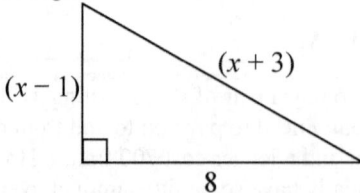

The Pythagoras theorem states that '*the square on the hypotenuse is equal to the sum of the squares on the two arms of any right angle triangle*'.

18. During an auction sale a phone was bought for 7200 francs, which is just $\frac{3}{4}$ of the regular price. Find the regular price of the phone and the amount gained by

19. Bame has 5000 francs and needs to buy a mathematics textbook which cost 3500 francs and exercise books which cost 300 francs each. How many exercise books will he be able to buy with the amount he has?

16.12 Inequalities and Inequations

Meaning of an Inequality

A statement that two real quantities or expressions are not equal is called an **inequality**.

$<$ means 'is less than'

$>$ means 'is greater than'

$\leq$ means 'is less than or equal to'

$\geq$ means 'is greater than or equal to'

Ordering

$\forall a, b, c \in \mathbb{R}$,

1. Either $a < b$ or $a > b$ or $a = b$.
2. $a < b \Leftrightarrow b > a$.
3. If $a < b$ and $b < c$ then $a < c$. This is known as the **transitive property** of inequalities.
4. If $a < b$ and $c > 0$ then $ac < bc$ and if $a < b$ and $c < 0$ then $ac > bc$.

An inequality, which involves an unknown, is called an **inequation**.

16.13   Representation of inequalities

On a real number line, if $a < b$ then the point a is to the left of the point b. When representing inequalities on a number line an open circle o is used when the boundary point is not included and a fill-in circle ● is used when the boundary point is included.

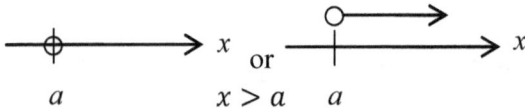

The boundary point a is not included.

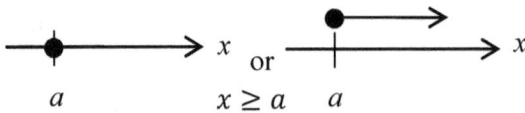

The boundary point a is included.

 Exercise 16:9

1. Represent each of the following on a number line.
   (a) $(-2, 0)$ (b) $]-4, 1[$ (c) $[-3, 1]$
   (d) $\{x: 3 < x < 8, x \in \mathbb{R}\}$ (e) $\{x: 4 \leq x \leq 8\}$
2. Name the inequality represented on the following number lines.

   (a)  (b)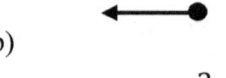

16.14   Building up Inequations

Inequations are built in the same way in which equations are built but for the fact that the inequality sign is used instead of the equal to sign.

Module 9, Topic 16: Simple Linear Equations and Inequalities

Example

1. The product of 2 and a number is at least 8.

    Solution
    Let the number be n. Then $2n \geq 8$.

2. When 5 is added to a number the result is greater than 9.

    Solution
    Let the number be x. Then $x + 5 > 9$.

3. When an amount of money is increased by 70 FCFA the total amount is at most 200 FCFA.

    Solution
    Let the amount of money be m. Then $m + 70 \leq 200$.

4. Four girls shared a number of oranges and each had at least 8 oranges.

    Solution
    Let the basket contain b oranges. Then $\frac{b}{4} \geq 8$.

Exercise 16:10

Build inequations from the following situations
1. When a number is increased by 8 the result is equal to or less than 14.
2. When 40, is divided by a certain number, the result is greater than 4.
3. Thrice the sum of a number and 7 is not more than 27.
4. The product of a number and 4 is at least 20.
5. 60 % of a number decreased by 10 is less than 12.
6. When half of a number is subtracted from three times the number, the difference is less than the number increased by 6.
7. Twice the sum of a number and 7 is at most 12 more than the number.
8. Three times the sum of a number and 5 is less than four times the number increased by 2.
9. Thirty minus five times a certain number is at least 4.
10. Three boys earn at most 600,000 FCFA. Ambe earns 20,000 FCFA less than Anye and Ndoh earns twice as much as Ambe.

16.15 Laws of Inequalities

Addition and Subtraction of Inequalities

 Investigative Activity

1. Write down any numerical inequality you like such as $8 > 6$.
   (a) Add a positive number such as $+2$ to both sides of the inequality.
   (b) Is the inequality still true?
   (c) What conclusion do you draw about adding a positive number to both sides of an inequality?
2. Write down another numerical inequality again such as $8 > 6$.
   (a) Add a negative number such as -4 to both sides of the inequality.
   (b) Is the inequality still true?
   (c) What conclusion do you draw about adding a negative number to both sides of an inequality?
3. Write down another numerical inequality.
   (a) Subtract a positive number such as $+4$ to both sides of the inequality.
   (b) Is the inequality still true?
   (c) What conclusion do you draw about adding a negative number to both sides of an inequality?
4. Write down another numerical inequality.
   (a) Subtract a negative number such as -4 from both sides of the inequality.
   (b) Is the inequality still true?
   (c) What conclusion do you draw about subtracting a negative number from both sides of an inequality?
5. What conclusion do you draw about adding any number to or subtracting any number from both sides of an inequality?

From the above investigations, we can see that:

If the same real quantity is added to or subtracted from both sides of an inequality, the inequality remains valid.

Module 9, Topic 16: Simple Linear Equations and Inequalities

Multiplication and Division of Inequalities

 Investigative Activity

1. Write down any numerical inequality you like such as $8 > 6$.
   (a) Multiply both sides of the inequality by a positive number such as $+2$.
   (b) Is the inequality still true?
   (c) What conclusion do you draw about multiplying both sides of an inequality by a positive number?
2. Write down any numerical inequality such as $8 > 6$.
   (a) Multiply both sides of the inequality by a negative number such as -2.
   (b) Is the inequality still true?
   (c) What conclusion do you draw about multiplying both sides of an inequality by a negative number?
3. Write down any numerical inequality such as $8 > 6$.
   (a) Divide both sides of the inequality by a positive number such as 2.
   (b) Is the inequality still true?
   (c) What conclusion do you draw about dividing both sides of an inequality by a positive number?
4. Write down any numerical inequality such as $8 > 6$.
   (a) Divide both sides of the inequality by a negative number such as -2.
   (b) Is the inequality still true?
   (c) What conclusion do you draw about dividing both sides of an inequality by a negative number?
5. What conclusion do you draw about multiplying or dividing both sides of an inequality by a positive number?
6. What conclusion do you draw about multiplying or dividing both sides of an inequality by a negative number?

From the above investigations, we can see that:

> 1. *If we multiply or divide both sides of an inequality by any positive real quantity, the inequality remains valid.*

But,

> 2. *If we multiply or divide both sides of an inequality by any negative real quantity, the inequality changes sense from $<$ to $>$ or $\leq$ to $\geq$ and vice versa.*

16.16 Solving Inequations

Inequations are solved in the same way as equations but for the fact that when multiplying or dividing both sides by a negative number, the inequality sign changes from < to > or from ≤ to ≥ and vice versa. We shall limit ourselves at this level to solving only simple linear inequations.

> **Example**

Solve the following inequalities and represent the solution on a number line.
1. $n+5>9$
2. $y+7<12$
3. $x-3 \geq -1$
4. $x-8 \leq 3$
5. $2x \geq 8$
6. $-32x < 64$
7. $\dfrac{x}{5} \leq 3$
8. $-\dfrac{x}{3} > 4$
9. $2x+5 \leq 13$
10. $\dfrac{2}{3}n+5 > 11$

Solution

1. $n+5>9$
   $\Rightarrow n>4$

2. $y+7<12$
   $\Rightarrow y<5$

3. $x-3 \geq -1$
   $\Rightarrow x \geq 2$

4. $x-8 \leq 3$
   $\Rightarrow x \leq 11$

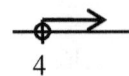

5. $2x \geq 8$
   $\Rightarrow x \geq 4$

6. $-32x < 64$
   $\Rightarrow x > -2$

7. $\dfrac{x}{5} \leq 3$
   $\Rightarrow x \leq 15$

8. $-\dfrac{x}{3} > 4$
   $\Rightarrow x < -12$

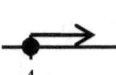

9. $2x+5 \leq 13$
   $\Rightarrow 2x \leq 8$
   $x \leq 4$

10. $\dfrac{2}{3}n+5 > 11$
    $\Rightarrow \dfrac{2}{3}n > 6$
    $2n > 18$
    $n > 9$

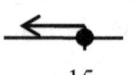

Module 9, Topic 16: Simple Linear Equations and Inequalities

 Exercise 16:11

Solve the following inequalities and represent your solution on a real number line.
1. $2x < 4$
2. $2 + x > 11$
3. $16 \leq 2x + 8$
4. $2y + 4 \geq 12$
5. $7x - 5 \leq 2x + 11$
6. $2(n + 5) > 18$
7. $\frac{2x-5}{3} < 25$
8. $5x > 40 - 3x$
9. $4y + 5 \leq 5y - 30$
10. $3t + 10 < 2t + 20$
11. $5u < 2u + 27$
12. $2x \geq 90 - 7x$
13. $10x - 11 > 8x$
14. $18 - 5x \leq x$
15. $13y \leq 15 + 3y$
16. $100 + 3\frac{1}{2}x > 23\frac{1}{2}x$
17. $9x \geq 16x - 105$
18. $3p + 3 < p + 8$
19. $5m + 10 > 16m - 12$
20. $20 \geq 3x - 1$
21. $0.6x - 0.4 \geq 0.4x + 0.6$

 Multiple Choice Exercise 16

1. The letter x is an unknown in:
   [A] $3x + 5 = 0$
   [B] $x^2 + x + 5$
   [C] $y = x^2 + x + 5$
   [D] $y = 3x + 5$

2. The letter x is a variable in:
   [A] $3x + 5 = 0$
   [B] $2x^2 + x + 5$
   [C] $(2x + 1)(x + 1)$
   [D] $y = 3x + 5$

3. The additive inverse of $\frac{7}{2}$ is:
   [A] $\frac{7}{2}$
   [B] $\frac{2}{7}$
   [C] $-\frac{7}{2}$
   [D] $-\frac{2}{7}$

4. The additive inverse of -14 is:
   [A] -14
   [B] 14
   [C] $\frac{1}{14}$
   [D] $-\frac{1}{14}$

5. The multiplicative inverse of 12 is:
   [A] -12
   [B] 12
   [C] $-\frac{1}{12}$
   [D] $\frac{1}{12}$

6. The multiplicative inverse of $\frac{2}{15}$ is:

[A] $-\dfrac{2}{15}$ [B] $\dfrac{15}{2}$ [C] $\dfrac{2}{15}$ [D] $-\dfrac{15}{2}$

7. Given that $8(x+8) = 40$. The statement, which best interprets this equation is:
   [A] Eight times the sum of a number and 8 is 40
   [B] Eight times a number less than eight is 40
   [C] Eight less than a number is 40
   [D] The product of eight and a number is 40

8. The root of the equation $3x + 4x = 42$ is:
   [A] $x = 4$ [B] $x = 6$ [C] $x = 8$ [D] $x = 7$

9. The root of the equation $3n + 14 = 47$ is:
   [A] $n = 8$ [B] $n = 9$ [C] $n = 10$ [D] $n = 11$

10. The root of the equation $6y - 48 = 2y$ is:
    [A] $y = 8$ [B] $n = 9$ [C] $n = 10$ [D] $n = 12$

11. The value of x for which $2x + 8 = 0$ is:
    [A] -4 [B] 4 [C] -6 [D] 6

12. Given that $33 = 6y + 3$. The value of y must be:
    [A] 3 [B] 5 [C] 6 [D] 8

13. If $3x - 7 = 10$, then the value of x is:
    [A] $\dfrac{3}{17}$ [B] $-\dfrac{17}{3}$ [C] $\dfrac{17}{3}$ [D] $-\dfrac{3}{17}$

14. If $6x + 4 = -20$, then the value of x is:
    [A] $-\dfrac{3}{8}$ [B] $\dfrac{8}{3}$ [C] -4 [D] 4

15. The value of x in the equation $5x + 1 = 31$ is:
    [A] 1 [B] 5 [C] 25 [D] 6

16. The value of y, which satisfies the equation, $4(y - 4) = 20$ is:
    [A] 1 [B] 24 [C] 6 [D] 9

17. The solution of $5(x - 4) - 4(x + 1) = 0$ is:
    [A] 16 [B] -24 [C] 24 [D] -16

18. The root of the equation $6(x - 4) + 3(x + 7) = 3$ is:
    [A] $\dfrac{3}{2}$ [B] $\dfrac{1}{3}$ [C] $\dfrac{1}{2}$ [D] $\dfrac{2}{3}$

19. $\dfrac{x-2}{3} = 8$, only if x is equal to:
    [A] 26 [B] 22 [C] 24 [D] 19

20. When $\dfrac{x}{4} = \dfrac{5}{2}$, the value of x is:
    [A] 2 [B] 4 [C] 5 [D] 10

21. Given that $\dfrac{1}{3}(x + 1) = 6$ the value of x is:
    [A] 19 [B] 17 [C] 5 [D] 3

22. Given that $\dfrac{2}{x} = \dfrac{3}{6}$, the value of x is:
    [A] 6 [B] 4 [C] 3 [D] 2

Module 9, Topic 16: Simple Linear Equations and Inequalities

23. The only condition for which $\dfrac{5}{x+1}$ is equal to 4 is that:

    [A] $x = 4$ [B] $x = 8$ [C] $x = \dfrac{1}{8}$ [D] $x = \dfrac{1}{4}$

24. Given that, $\dfrac{3-2y}{4} = \dfrac{2y}{6}$ the value of y is:

    [A] $\dfrac{10}{9}$ [B] $\dfrac{9}{10}$ [C] 3   [D] −3

25. The root of the equation $\dfrac{2x+7}{6} + \dfrac{x-5}{3} = 0$ is:

    [A] $x = -\dfrac{3}{5}$ [B] $x = \dfrac{3}{4}$ [C] $x = \dfrac{1}{4}$ [D] $x = \dfrac{2}{5}$

26. Given that $\dfrac{3x-2}{6} - \dfrac{2x+7}{9} = 2$. x is equal to:

    [A] $x = -\dfrac{36}{5}$ [B] $x = \dfrac{56}{5}$ [C] $x = -\dfrac{16}{5}$ [D] $x = \dfrac{16}{5}$

27. Using the relation $C = \dfrac{5}{9}(F - 32)$, the value of F when $C = 40$ is:

    [A] 67   [B] 77   [C] 81   [D] 104

28. The value of t, which satisfies $\dfrac{3t}{4} + \dfrac{1}{3}(21 - t) = 11$ is:

    [A] $9\dfrac{3}{5}$ [B] $3\dfrac{9}{13}$ [C] 5   [D] $\dfrac{9}{13}$

29. If $8x - 4 = 6x - 10$, the value of $5x$ is:

    [A] 7   [B] −15   [C] −3   [D] 3

30. The value of x, which satisfies the equation, $5(x-7) = 7x - 5$ is:

    [A] $x = 6$ [B] $x = -30$ [C] $x = -15$ [D] $x = -6$

31. If $2(x+1) = 4x + 3$, x equals:

    [A] 2   [B] −2   [C] $\dfrac{1}{2}$ [D] $-\dfrac{1}{2}$

32. The value of $3(p+7)$ for which $6p + 5 = 4p + 11$ is:

    [A] 15   [B] 20   [C] 25   [D] 30

33. The equation $\dfrac{2}{3}(x+5) = \dfrac{1}{4}(5x - 3)$ has root:

    [A] $1\dfrac{1}{7}$ [B] 7   [C] 3   [D] $4\dfrac{3}{7}$

34. Given that $\dfrac{m}{3} + \dfrac{1}{2} = \dfrac{3}{4} + \dfrac{m}{4}$, the value of m is:

    [A] −3   [B] −2   [C] 2   [D] 3

35. The only condition for $0.6x - 0.4$ to be equal to $1.2x + 0.8$ is that:

    [A] $x = -2$ [B] $x = 2$ [C] $x = -0.5$ [D] $x = -0.5$

36. Given that $0.9n - 0.7 = 0.3n - 0.1$, then n must be:

[A] 4 [B] 3 [C] 2 [D] 1

37. A man is 23 years older than his son is this year. Given that his son will be 12 years old in ten years' time, the man will be:
    [A] 25 years [B] 35 years [C] 33 years [D] 37 years

38. Abe bought two packets of sugar at x FRS each and four tins of milk at 400 FRS each. If the total cost is 3400 FRS, the price of a packet of sugar is:
    [A] 400 FRS [B] 1800 FRS [C] 900 FRS [D] 1600 FRS

39. 8 more than thrice a number is 35. The number is:
    [A] 27 [B] 43 [C] 9 [D] 14.3

40. A trader bought 400 liters of palm oil and sold $8x$ liters. If 160 liters are left, the number of liters he sold is:
    [A] 30 [B] 240 [C] 5 [D] 20

41. Given that $x = 3$. The number, which we can add to $12x$ to make 57, is:
    [A] 21 [B] $\dfrac{19}{4}$ [C] 45 [D] 33

42. The sum of three consecutive numbers is 42. The largest of the numbers is:
    [A] 13 [B] 14 [C] 15 [D] 16

43. Three quarters of a certain number is 15. The number is:
    [A] 20 [B] 16 [C] 15 [D] 12

44. The ratio of $2x$ to $x + 1$ is $5 : 3$ only if x is:
    [A] $x = 5$ [B] $x = 4$ [C] $x = 3$ [D] $x = 2$

45. The ages of two people are in the ratio $5: 9$. If the elder is 8 years older, the age of the younger is:
    [A] 45 years [B] 10 years [C] 8 years [D] 2 years

46. The solution of the inequality $3m + 3 > 9$ is:
    [A] $m > 2$ [B] $m > 3$ [C] $m > 4$ [D] $m > 6$

47. If x is positive, the range of values of x for which $4 + 3x < 10$ is:
    [A] $0 < x < 2$ [B] $x < 2$ [C] $1 < x < 2$ [D] $0 > x > 2$

48. The inequality $\frac{1}{3}(2x - 1) < 5$ is true only if:
    [A] $x < -5$ [B] $x > -5$ [C] $x < 7$ [D] $x > 8$

49. p and q are two positive real numbers such that $p > 2q$. The inequality which is not true is:
    [A] $-p < -2q$ [B] $-p > 2q$ [C] $-p < 2q$ [D] $-q < \frac{1}{2}p$

50. The solution of the inequality $y - 3 < \frac{y}{3}$ is:
    [A] $y > -\frac{9}{2}$ [B] $y < 3$ [C] $y > 4$ [D] $y < \frac{9}{2}$

51. The solution of the inequality $3x - 8 \geq 5x$ is:
    [A] $x \geq 4$ [B] $x \geq 1$ [C] $x \leq -4$ [D] $x \leq -1$

52. The range of values of x which satisfy the inequality $2x + 3 < 5x$ is:
    [A] $x > 1$ [B] $x < \frac{3}{7}$ [C] $x > \frac{3}{7}$ [D] $x > -1$

53. Nfor had x oranges. He ate 2 and shared the remainder equally with Ngala. In terms of x, the inequality which represents the information that Ngala's share is at least 5 oranges is:

Module 9, Topic 16: Simple Linear Equations and Inequalities

[A] $\frac{x}{2} - 2 \leq 5$ [B] $\frac{x}{2} - 2 \geq 5$ [C] $\frac{x+2}{2} \geq 5$ [D] $\frac{x-2}{2} \leq 5$

54. The smallest whole number which satisfies the inequality $9 - 2x < 5x - 12$ is:
   [A] 1   [B] 2   [C] 3   [D] 4

55. The range of values of x for which $\frac{1}{2}(4x + 2) - (x - 5) \leq \frac{1}{4}(3x - 1)$ is:
   [A] $x \geq 25$ [B] $x \leq 25$ [C] $x \geq -25$ [D] $x \leq -25$

56. In the following figure, the solution to the inequality $\frac{x}{3} - \frac{x-3}{2} < 1$ is represented by the line:

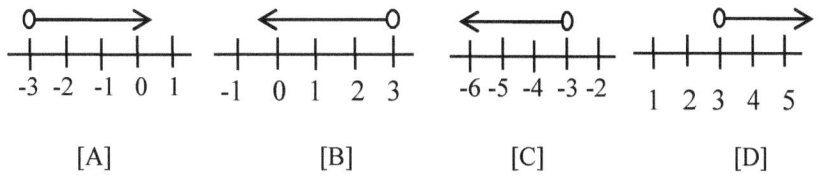

[A]   [B]   [C]   [D]

Answers to Structural Exercises

Review Exercise 1:1

1. $\mathbb{N}$ 2. $\mathbb{N}^*$ 3. $\{0, 1, 2, 3, 4\}, \{1, 2, 3, 4, 5\}$ 4. 76, 77, 79, 98

5. (a) >   (b) <   6. $56 < 63 < 68 < 74$

7. Index, exponent, power or logarithm; base

8. (a) 3^9 (b) 6^5 9. (a) 9×9 (b) $5 \times 5 \times 5$ 10. 10^7

11. (a) 81   (b) 64   12. (a) $\{1, 3, 5, 7, 9\}$ (b) $\{2, 4, 6, 8, 10\}$

13. (a) 103   (b) 124   14. $\mathbb{Z} = \{..., -3, -2, -1, 0, 1, 2, 3, ...\}$

Skill Building Exercise 1:1

1. 64   2. 64   3. 72   4. 675   5. 729   6. 250

7. 972   8. $\frac{1}{169}$ 9. 18   10. 15   11. 288   12. 128

13. $\frac{10}{21}$ 14. $\frac{5}{3}$ 15. -1715 16. $\frac{1280}{186}$ 17. $\frac{8}{405}$ 18. $\frac{1875}{2}$

19. $\frac{16}{15}$ 20. 5   21. $\frac{1}{3}$ 22. $\frac{8}{15}$ 23. $\frac{864}{216}$ 24. 3

Skill Building Exercise 1:2

1. 2^{24} 2. 3^{-12} 3. 1   4. 13^{10} 5. 1   6. -2^{24} 7. -2^{15}

8. -3^{-8} 9. -3^8 10. $2^{10} \times 3^{-2} \times 4^{-3}$ 11. $12 \times 5^{-3} \times 9^{-1} \times 4^{-1}$

Exercise 2:1

1. (a) 13^8 (b) 31^6 (c) 17^{10} (d) 20^4 2. (a) 4^6 (b) 7^{12} (c) 1^{15} (d) 0^{18}

3. (a) 27   (b) 25   (c) 90   (d) 64   4. (a) 24   (b) 32   (c) 80   (d) 35

5. (a) 8   (b) 12   (c) 18   (d) 14   6. 2

Exercise 2:2

1. $\frac{2}{3}$ 2. $\frac{6}{7}$ 3. $\frac{3}{2}$ 4. $\frac{3}{4}$ 5. $\frac{3}{7}$ 6. $\frac{5}{3}$ 7. $\frac{7}{8}$ 8. $\frac{5}{3}$ 9. $\frac{9}{4}$ 10. $\frac{8}{3}$ 11. $\frac{12}{7}$ 12. $\frac{3}{2}$

Exercise 2:2

Answers to Structural Exercises

1. (a) 10, 12, 14 (b) 27, 33, 39 (c) 25, 36, 49 (d) 20, 23, 26
   (e) 10, 0, -10 (f) 11, 16, 22 (g) $1, \frac{1}{3}, \frac{1}{9}$ (h) 13, 16, 19
   (i) 42, 52, 62 (j) 25, 21, 17 (k) 17, 27, 24 (l) $-16, -20, -24$
   (m) 33, 45, 59 (n) 19, 23, 27 (o) 28, 36, 45   2. 15, 21, 28

Skill Building Exercise 3:1

1. (a) 1000 (b) 500 (c) 9500 (d) 5100 (e) 12100
   (f) 350 (g) 470 (h) 47400   2. (a) 3 (b) 12 (c) 4 (d) 6
3. (a) 1 (b) 15 (c) 1 (d) 4

Skill Building Exercise 3:2

1. (a) 500 (b) 50000 (c) 30 (d) 0.8 (e) 54 (f) 24
2. (a) 3 (b) 10 (c) 2 (d) 0.1 (e) 20 (f) 0.08

Skill Building Exercise 3:3

1.

	Number of significant figures			
	1	2	3	4
(a)	0.0	0.01	0.007	0.0068
(b)	2	2.0	2.01	2.007
(c)	5	4.7	4.70	4.698
(d)	1	1.0	1.01	1.006

2.

	Number of decimal places			
	1	2	3	4
a	0.0	0.01	0.007	0.0068
b	2.0	2.01	2.007	2.0068
c	4.6	4.70	4.698	4.6977
d	1.0	1.01	1.006	1.0061

3. (a) 0.004 (b) 0.457 (c) 0.505   4. (a) 14.91 (b) 23.11 (c) 6.04
5. (a) 0.025 (b) 4.02   6. (a) 54 (b) 59 (c) 54   7. (a) 0.0 (b) 5 (c) 0.2

Skill Building Exercise 3:4

(a) 5×10^3 (b) 4.8×10^2 (c) 1.02×10^4 (d) 7.0×10^5 (e) 3.2×10^{-3} (f) 7.3×10^{-5} (g) 9.25×10^{-1} (h) 1.1×10^{-4} (i) 0.56×10^4 (j) 3×10^{-5} (k) 1.96×10^{-3} (l) 3.4×10^{-4}

Skill Building Exercise 3:5

1. (a) 91.92×10^6 (b) 3.5×10^1 (c) 5.4×10^{-3} (d) 2.0×10^1
 (e) 3.5×10^2 (f) 6.0×10^{-2} (g) 4.0×10^4 (h) 9.687×10^2
 (j) 3.66×10^{-2} (j) 0.13 (k) 6.0×10^{-2} (l) 4370×10^{-3}
2. 3.0×10^4 3. 1.7×10^{24}
4. (a) 9.73×10^{-1} (b) 1.0 (c) 1.0 5. (a) 142×10^{-4} (b) 0.01 (c) 0.014

Exercise 4:1

1. 9900 Frs. 2. 17 days 3. 10800 Frs.
4. (a) 20 hours (b) 40 hours (c) 30 hours
5. (a) 40 km (b) 120 km (c) 480 km
6. (a) 120 Frs. (b) 2000 Frs. (c) 2400 Frs.
7. (a) 9800 question papers (b) 17 teachers

Exercise 4:2

1. (a) $1\frac{1}{4}$ days (b) $12\frac{1}{2}$ days (c) 25 days
2. (a) 6 minutes (b) 3 minutes (c) 2 minutes
3. (a) 6 minutes (b) 2 minutes (c) 3 minutes
4. 4 days 5. 5 women 6. 6 guavas 7. 16 hours

Exercise 4:3

1. 24% 2. $16\frac{2}{3}$% 3. 29.2% 4. 10%
5. 12.5% 6. 22.8% 7. 10% 8. $3\frac{1}{3}$%

Exercise 4:4

Answers to Structural Exercises

1. (a) 180,000 FCFA (b) 9,600 FCFA (c) 60,000 FCFA
2. 1,000,000 FCFA   3. 9.9%   4. 4years
5(a) 3%   (b) 200,000 FCFA   (c) 2years

Exercise 4:5
1. 3,340,810 FCFA   2. 61,800 FCFA   3. 5,701,440 FCFA
4. 331013 FCFA   5. 4 years   6. 562,432 FCFA   7. 374,562 FCFA

Exercise 4:6
1. 779280 FCFA   2. 2454861.6 FCFA   3. £ 2251

Skill Building Exercise 5:1
(a) 50   (b) 20   (c) 24   (d) 33   (e) 35
(f) 9   (g) 12   (h) 18   (i) 14   (j) 15

Exercise 5:1
1. irrational 2. rational 3. irrational 4. irrational 5. rational
6. irrational 7. rational 8. rational 9. Irrational 10. irrational

Exercise 5:2
1.

   A: $-\dfrac{10}{3}$, B: -2.5, C: 0, D: 2, E: $\dfrac{13}{4}$, F: $\sqrt{13}$
   (number line from -4 to 5)

2. (a) 2 and 6   (b) -5 and -1 (c) -1 and 2
   (d) 2 and 6   (e) 2 and 6   (f) -5 and -1

3. A, C, E, B, D.

Exercise 5:3
1. Rational: (a), (b) and (e)   Irrational: (c), (d) and (f)

2.

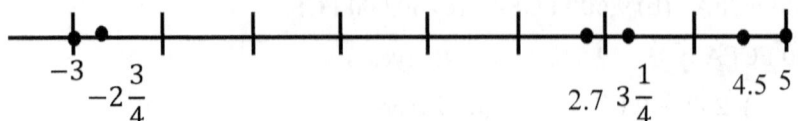

3. $-10, -9, -8, -7, -6, -5, -4, -3, -2, -1, 0, 1, 2, 3, 4, 5, 6, 7, 8, 9, 10$

4. Infinite because a rational number can always be found between any two rational numbers.

6. (a) $\frac{2}{5} > \frac{3}{8}$ (b) $-\frac{1}{2} < -\frac{1}{3}$ (c) $\frac{1}{2} > \frac{1}{3}$
(d) $-3.54 < 3.549$ (e) $0.57 < \frac{4}{7}$ (f) $-\frac{2}{3} < -\frac{4}{5}$
(g) $\frac{2}{3} > \frac{4}{5}$ (h) $-\frac{7}{9} > -0.779$ (i) $\frac{11}{13} < \frac{23}{25}$

7.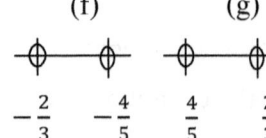

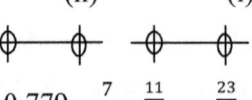

8. d

	N	Z	Q	Q'	R
			$\frac{2}{3}$	$\frac{2}{3}$	$\frac{2}{3}$
			-5.3	-5.3	-5.3
			$1\frac{1}{2}$	$1\frac{1}{2}$	$1\frac{1}{2}$
			$-\frac{4}{11}$	$-\frac{4}{11}$	$-\frac{4}{11}$
				π	π
				$-\sqrt{3}$	$-\sqrt{3}$

9. (a)   N   (b) Q and R   (c) Q and Q'   (d) Q and Q'

Answers to Structural Exercises

Exercise 5:4

1. (a) (b) (c) (d) (e)

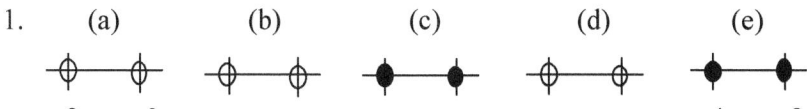

2. (a) $[-5,3]$ (b) $(4,8)$ (c) $(-4,2)$ (d) $(5,11)$ (e) $x > 5$ (f) $x \leq -2$

3. (a) closed (b) Half opened, half closed (c) Half opened, half closed
   (d) Opened (e) Closed (f) Opened (g) Closed

4. (a) (b) (c) (d) (e)

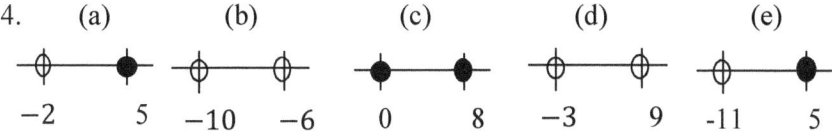

5. P is closed; both 17 and 19 are included.

    Q is closed; both 1 and 3 are included
    R is half opened, half closed, -1 is excluded but 2 is included.
    S is half opened, half closed, -2 is included but 2 is excluded.
    T is closed; both 2 and 5 are included.
    U is opened; both -2 and 0 are excluded.
    V is opened; both 5 and 8 are excluded.
    W is opened; both -7 and -2 are excluded.

6.   P   Q   R   S

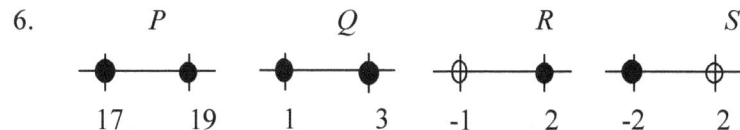

    T   U   V   W

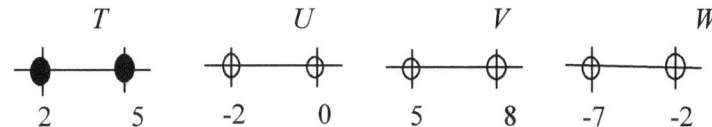

7. (a) $(-2,2]$ or $]-2,2]$ (b) $98 \leq x \leq 101$ (c) $]2,5[$ or $2 < x < 5$
   (d) $[1,3[$ or $[1,3)$ (e) $(-6,-1)$ or $-6 < x < -1$
   (f) $]14,19]$ or $14 < x \leq 19$ (g) $[0,7[$ or $0 \leq x < 7$
   (h) $[-3,1)$ or $-3 \leq x < 1$ (i) $(11,13]$ or $11 < x \leq 3$

Exercise 6:1

1. (a) 5 (b) 15 (c) 10 (d) 15 (e) 15

Competency Base Mathematics for Secondary Schools Book 2

2. (a) 3cm   (b) 5.5cm   3. (a)   (b)
4. (a) 5.5cm   (b) 3cm   5. 25.5   6. 23.9
7. (a) 3   (b) 6.5   (c) 3.5   (d) 10   (e) 7

Exercise 7:1

1. $x = 60°$, $y = 60°$ 2. $b = 60°$ 3. $x = 5$, $y = 35°$
4. $x = 70°$ 5. $s = 333°$

Exercise 7:2

(1) $89°$ (2) 9   (3) $40°$ (4) 5   (5) 6   (6) 12   (7) 12   (8) 18
(9) $144°$ (10) (a) Yes, 10   (b) Yes, 9   (c) No.
(11) (a) Yes, 18   (b) No   (c) Yes   (12) 10   (13) $108°$
(14) 12, $\angle ACD = 135°$

Exercise 8:1

(a) $x = 70°$ (b) $x = 120°$, $y = 40°$ (c) $x = 20°$, $y = 136°$
(d) $x = 80°$, $y = 40°$ (e) $x = 72°$, $y = 54°$ (f) $x = 75°$, $y = 30°$ (g) $p = 115°$
                               (h) $y = 80°$ (i) $x = 70°$, $y = 50°$

Exercise 8:2

1. (a) 13 cm   (b) 10 cm   (c) 9 cm   (d) 12 cm   (e) 9 cm
   (f) 17 cm   (g) 45 cm   (h) 16 cm   (i) 2 cm   (j) 8 cm
2. 4 cm   3. 41 cm   4. 18 cm   5. 12 cm, 16 cm   6. a, c, e, h

Exercise 8:3

	Number	Name of pairs	Matching sides
1.	1	PQT and RST	PQ and RS, PT and RT, QT and TS
2.	2	$\triangle ABC \equiv \triangle AED$ $\triangle ABD \equiv \triangle AED$	$AB = AE$, $AC = AD$, $BC = DE$, $BD = CE$
3.	1	$\triangle XOY \equiv \triangle XOZ$	XO is common, $OY = OZ$, $XY = XZ$
4.	1	$\triangle POM \equiv \triangle QLN$	$PO = QL$, $PM = QN$, $OM = LN$
5.	4	$\triangle POQ \equiv \triangle ROS$ $\triangle ROQ \equiv \triangle POS$ $\triangle QRS \equiv \triangle PQS$ $\triangle PRS \equiv \triangle PRQ$	$PQ = RS$, $PT = TR$, $QT = TS$, $OQ = OS$, $PI = OR$
6.	1	$\triangle POQ \equiv \triangle RTS$	$PQ = RS$, $PT = TR$, $QT = TS$, $TQ = TS$, $PI = TR$
7.		$\triangle ZWX \equiv \triangle YXW$ $\triangle ZWO \equiv \triangle YXO$ $\triangle ZWY \equiv \triangle YXZ$	$WZ = XY$, $WY = XZ$, $WO = XO$, $ZO = YO$, WX is common, ZY is common

8. 2 sets; {$\triangle ADF, \triangle DBE, \triangle CFE, \triangle DEF$}, {$\triangle IDG, \triangle FHI, \triangle EGH, \triangle GHI$}

Answers to Structural Exercises

Exercise 9:1

1. (i) 6 (ii) 7
2.

Name of Prism	Number of		
	Faces	Edges	Vertices
Triangular prism	5	9	6
Cube	6	12	8
Cuboid	6	12	8
Pentagonal prism	7	15	10
Hexagonal prism	8	18	12

Exercise 9:2

1. 8400 cm^2 2. 1440 cm^3 3. 40 cm 4. 13 m 5. 10 m
6. 0.003 m^3 7. $144000\ l$ 8. $10000\ l$ 9. $60000\ l$
10. (a) 8 cm^3 (b) 24 cm^2 11. 192
12. (a) 960 cm^3 (b) 784.4 cm^2 13. (a) 277 cm^3 (b) 295.4 cm^2

Exercise 9:3

1. 12320 cm^2 2. 528 cm^2 3. 5 cm
4. (c) 785.7 cm^2 and 2514.3 cm^2 (d) 1571.4 cm^3 and 9428.6 cm^3
5. 63.25 m^2 6. 1.75 cm 7. 1078 cm^3 8. 3 cm 9. 22000 cm^3 10. $308\ l$ 11. 127.3 cm 12. (a) $528\ cm^2$ (b) $2304\ cm^3$

Exercise 10

1. (a) 528 cm^2 (b) 768 cm^3 2. (a) 216 cm^2 (b) 192 cm^3 3. 4700 cm^2

Exercise 11:1

1. (a) 2 km (b) 15 km (c) 25.2 km
2. (a) 1:1200 (b) 10 cm (c) 8.4 m
3. Answers may vary. but the scale must be one that allows 3m to be represented by 20 cm. One good scale may be 6 cm : 1 m.
6. Corresponding sides are in a common ratio and corresponding angles are equal.
7. (a) 30 km (b) 1.6 cm (c) 1 cm to 1 km or 1:1,000,000

8. All right isosceles triangles are similar because the corresponding angles are equal.
9. (a) Height = 8.2 cm, diameter = 2.6 cm. (b) 1 : 410   10. 41 cm

Exercise 11:1

1. Congruent   2. Not congruent   3. Congruent   4. Not congruent
5. Not congruent   6. Not congruent   7. Not congruent
8. $\triangle PTQ \equiv \triangle RTS$, $\triangle PQR \equiv \triangle SRQ$, ASA
9. $\triangle ABC \equiv \triangle AED$, $\triangle ABD \equiv \triangle AEC$, SAS
10. $\triangle XOY \equiv \triangle XOZ$, SAS
11. $\triangle POM \equiv \triangle QLN$, ASA
12. $\triangle POS \equiv \triangle ROQ$, SAS or ASA or SSS
    $\triangle POQ \equiv \triangle ROS$, ASA
    $\triangle PQS \equiv \triangle RSQ$, SAS
    $\triangle PRS \equiv \triangle RPQ$, SAS
13. $\triangle WOZ \equiv \triangle XOY$, ASA
    $\triangle WYZ \equiv \triangle XZY$, ASA
14. Congruent   15. Congruent   16. Congruent
17. Congruent   18. Not congruent   19. Congruent

Exercise 12:2

1. Pie chart to be drawn   2. (a) 6000 frs   (b) 750 frs   3. Histogram to be drawn   4. 40%   5. Histogram to be drawn   6. 14°   7. A bar chart to be drawn
8. (a) 100°   (b) 15   9. $w = 156°, x = 72, y = 16$ and $z = 24°$ 10. 105°

Exercise 13:1

1. 13   2. 1   3. 9   4. (a) 54kg   (b) 51.2kg   (c) 53.5kg
5. (a) 2.9   (b) 3   (c) 3.5   6. (a) 5   (b) 5.3   (c) 5
7. (a) 7   (b) 7.1   (c) 7   8. (a) 113.5   (b) 100   (c) 50
9. (a) 70   (b) 68   (c) 70   10. (a) 30   (b) 8 marks   (c) 6.1
11. (a) 2   (b) 2   (c) 2   12. (a) 1   (b) 3   (c) 2

Answers to Structural Exercises

Exercise 14:1
1. (a) S = {blue, red, green, white, black}   (b) E ={white, green}
2. HH, HT, TH, TT   3. S = {1,2,3,4,5,6}
4. (a) {2,3,5,7} (b) {2,4,6,8,10} (c) {1,3,5,7,9} (d) {1,4,9} (e) {3,6,9}
5. {A, E, A, I}   6. 12

Exercise 14:2
1. (a) $\frac{1}{6}$ (b) $\frac{1}{2}$ (c) $\frac{1}{2}$ (d) $\frac{2}{3}$ (e) $\frac{1}{3}$ 2. (a) $\frac{3}{8}$ (b) $\frac{1}{8}$ (c) 0 (d) $\frac{1}{8}$ 3. $\frac{1}{4}$ 4. $\frac{3}{10}$ 5. $\frac{2}{11}$ 6. $\frac{1}{3}$

Exercise 14:3
1. (a) $\frac{9}{35}$ (b) $\frac{23}{35}$ (c) $\frac{14}{35}$ 2. $\frac{11}{15}$ 3. $\frac{21}{25}$ 4. $\frac{3}{10}$ 5. $\frac{3}{7}$ 6. $\frac{1}{4}$

Exercise 15:1
1. $2x$ 2. $y+3$ 3. $p-5$ 4. $x+5$ 5. $x+8$
6. $y-9$ or $9-y$ 7. $2y$ 8. $\frac{1}{2}y$ 9. $p-10$ 10. $2q+7$ 11. $\frac{x}{2}$
12. $4p$ 13. $\frac{y}{4}$ 14. y^2 15. x^3 16. $3y-4$ 17. $\sqrt{2x}$
18. x^2+2 19. $15-y$ 20. $2y-8$ 21. $\frac{1}{2}x-7$ 22. $\frac{1}{2}x+7$

Exercise 15:2
1. 7   2. 1   3. 10 cm²   4. 15   5. 36   6. 88 cm³   7. 154 cm²

Exercise 15:3
1. (a) $6x, -2xy$ and $3y$ (b) px and $\frac{p}{y}$ (c) w and $5pz$ (d) $3pt, -\frac{8x}{y}$ and $\frac{1}{w}$
2. (a) 7   (b) −2   (c) 5   3. (a) pt (b) $\frac{x}{y}$ (c) pw

Exercise 15:4
1. $21x-17y$ 2. $s-t$ 3. $5p-22q$ 4. $-x-7$ 5. $u-3v$
6. $20a+10b$ 7. $2x+18y+z$ 8. b 9. $-99x-8z+88$ 10. $-pq+2rq$

Competency Base Mathematics for Secondary Schools Book 2

Exercise 15:5
1. (a) $8b$ (b) $3x^2$ (c) $20pq$ (d) $18ab^2$ (e) $24x^2y$
   (f) $28x^2y$ (g) $50ab$ (h) $30a^2$ (i) $20x^2y$ (j) $27u^2v^2$

2. (a) $2xy$ (b) $6p$ (c) $7x$ (d) $6x$ (e) $11b$ (f) $6p$
   (g) $\frac{7xy}{8}$ (h) $4u$ (i) $5y$ (j) $3m$ (k) $\frac{4p^2+4q^2}{pq}$ (l) x

Exercise 15:6
(a) $3x + 15$ (b) $4y + 8$ (c) $4x + xy$ (d) $3v + uv$ (e) $2x - 2$
(f) $5y - 15$ (g) $6p - 3pq$ (h) $4t - st$ (i) $10x - 2x^2$ (j) $5xy - 15y$

Exercise 16:1
1. $3 + x = 4$ 2. $7d = 42000$ 3. $y + 3 = 14$ 4. $w + 3w = 16$
5. $3(x + 6) = 33$ 6. $x + 20 = 3x$ 7. $x + 5 = 2x - 4$ 8. $12h = 120 - 36$
9. $60m = 450$ 10. $16960 + t = 17810$

Exercise 16:2
1. (a) 5 (b) 8   2.(a) 10 b) 17
3. (a) divide by 3 or multiply by $\frac{1}{3}$ (b) divide by 7 or multiply by $\frac{1}{7}$
   (c) multiply by 5   (d) multiply by   (e) divide by 0.05
   (f) multiply by 0.02   (g) subtract 5   (h) multiply by 13 and add 2

Exercise 16:3
1. $x = 6$ 2. $t = 7$ 3. $x = 10$ 4. $a = 8$ 5. $x = 7$ 6. $y = 10$ 7. $x = 9$ 8. $x = 4$ 9. $y = 4$ 10. $x = \frac{16}{5}$ 11. $n = \frac{13}{2}$ 12. $x = 40$

Exercise 16:4
1. $x = 8$ 2. $t = 3$ 3. $p = -6$ 4. $u = 42$ 5. $z = 36$ 6. $w = -33$ 7. $r = -27$ 8. $q = -6$

Exercise 16:5
1. $x = 2$ 2. $t = \frac{5}{2}$ 3. $q = 4$ 4. $p = 1$ 5. $a = 7$ 6. $x = 8$ 7. $d = 8$ 8. $y = \frac{9}{2}$
9. $u = 16$ 10. $m = 40$ 11. $n = 4$ 12. $x = 230$ 13. $k = 40$ 14. $z = 60$
15. $n = \frac{13}{2}$ 16. $u = 9$ 17. $x = 8$ 18. $x = 10$ 19. $t = \frac{11}{18}$ 20. $a = 3$
21. $u = 15$ 22. $y = \frac{3}{2}$ 23. $y = 35$ 24. $t = 10$

Answers to Structural Exercises

Exercise 16:6
1. $p = \frac{5}{2}$ 2. $m = 2$ 3. $y = 10$ 4. $x = 3$ 5. $x = 2$ 6. $x = 2$
7. $m = -42$ 8. $x = \frac{10}{3}$ 9. $t = -20$ 10. $x = 1$ 11. $p = -1$ 12. $x = 4$

Exercise 16:7
1. $x = 5$ 2. $p = -1$ 3. $x = \frac{2}{3}$ 4. $x = -\frac{10}{21}$ 5. $w = 123$ 6. $x = 76$ 7. $x = 40$
8. $x = 3$ 9. $y = \frac{7}{30}$ 10. $x = 5$

Integration exercise : You will buy 7 ledgers and be left with 500frs

Exercise 16:8
1. 13 cm   2. 8 mangoes   3. $y = 100$ 4. $x = 2$ 5. 4   6. $x = 5$
7. $d = 600$ 8. 7 years   9. 4 matches   10. 5   11. 10   12. 9
13. 7 minutes   14. $t = 850$ frs   15. $6000 + 3000n = 27000, n = 7$ 16. 9
17. $x = 7$ 18. 9600 frs, 2400 frs   19. 5 exercise books

Exercise 16:9

1. (a) 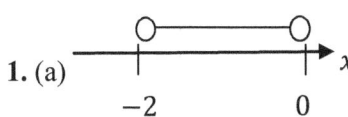 (b)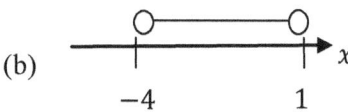

(c) 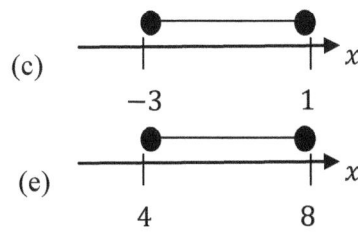 (d)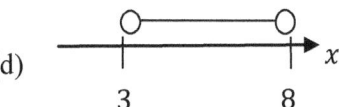

(e)

2. (a) $x > 5$ (b) $x \leq -2$

Exercise 16:10
1. $x + 8 \leq 14$ 2. $\frac{40}{x} > 4$ 3. $3(x + 7) \leq 27$ 4. $4x \geq 20$ 5. $\frac{60x}{100} - 10 < 12$
6. $3x - \frac{x}{2} < x + 6$ 7. $2(x + 7) \leq 12 + x$ 8. $3(x + 5) < 4(x + 2)$
9. $30 - 5x \geq 4$ 10. $4x - 60000 \leq 600000$

Exercise 16:11

1. $x < 2$
2. $x > 9$
3. $x \geq 4$
4. $y \geq 4$
5. $x \leq \dfrac{16}{5}$
6. $n > 4$
7. $x < 40$
8. $x > 5$
9. $y \geq 35$
10. $t < 10$
11. $u < 9$
12. $x \leq 10$
13. $x > \dfrac{11}{2}$
14. $x \geq 3$
15. $y \leq \dfrac{3}{2}$
16. $x < 5$
17. $x \leq 15$
18. $p < \dfrac{5}{2}$
19. $m < 2$
20. $x \leq 7$
21. $x \geq 5$

www.ingramcontent.com/pod-product-compliance
Lightning Source LLC
Chambersburg PA
CBHW071420180526
45170CB00001B/161